IN THE SHADOW OF SLAVERY

IN THE SHADOW OF SLAVERY

Africa's Botanical Legacy in the Atlantic World

JUDITH A. CARNEY AND
RICHARD NICHOLAS ROSOMOFF

UNIVERSITY OF CALIFORNIA PRESS

BERKELEY LOS ANGELES LONDON

University of California Press, one of the most distinguished
university presses in the United States, enriches lives around
the world by advancing scholarship in the humanities, social
sciences, and natural sciences. Its activities are supported by
the UC Press Foundation and by philanthropic contributions
from individuals and institutions. For more information,
visit www.ucpress.edu.

University of California Press
Berkeley and Los Angeles, California

University of California Press, Ltd.
London, England

Library of Congress Cataloging-in-Publication Data

Carney, Judith Ann.
    In the shadow of slavery : Africa's botanical legacy in
the Atlantic world / Judith A. Carney, Richard Nicholas
Rosomoff.
        p.   cm.
    Includes bibliographical references and index.
    ISBN 978-0-520-25750-4 (cloth : alk. paper)
    1. Blacks—Ethnobotany—America—History.
    2. Blacks—Ethnobotany—Africa—History.
3. Slaves—America—History.   4. Ethnobotany—
America—History.   5. Ethnobotany—Africa—History.
6. Plants, Edible—America—History.   7. Plants,
Edible—Africa—History.   8. Medicinal plants—
America—History.   9. Medicinal plants—Africa—
History.   10. America—Civilization—African influences
I. Rosomoff, Richard Nicholas, 1956–   II. Title.

E29.N3C38 2009
581.6'3097—dc22   2009015360

Manufactured in the United States of America

19 18 17 16 15 14 13 12 11
10 9 8 7 6 5 4 3 2 1

This book is printed on Cascades Enviro 100, a 100% post
consumer waste, recycled, de-inked fiber. FSC recycled
certified and processed chlorine free. It is acid free, Ecologo
certified, and manufactured by BioGas energy.

For Elaine Rosomoff
and in loving memory of Margaret G. Carney,
Louis J. Carney, and Hubert Rosomoff

CONTENTS

ILLUSTRATIONS

FIGURES

PLATES

Plates follow page 176.

ACKNOWLEDGMENTS

This book was generously supported by fellowships from the American Council of Learned Societies and the John Simon Guggenheim Memorial Foundation. The African Studies and Latin American Studies centers of the University of California, Los Angeles, provided crucial funding for travel. A residency at the Rockefeller Foundation Study and Conference Center in Bellagio, Italy, provided a tranquil and nurturing setting in which to develop some of the principal themes and to organize the book's outline.

Over the many years this book evolved, several scholars gave selflessly of their time and critical energies to review drafts of the manuscript and to enrich it with their invaluable commentary and expertise. Special gratitude must be extended to Chris Ehret, James C. McCann, James H. Sweet, Gail Wagner, and Kairn Kleiman for their close readings. Knowledge is a social process, and all who follow the path of scholarship depend on the encouragement, insights, and healthy critiques of other colleagues. Several scholars who deserve particular mention in the development of this book include Eugene Anderson, David P. Gamble, Kat Anderson, Jerry Handler, Stephen Bell, Walter Hawthorne, Paul Robbins, Sidney Mintz, Bruce Mouser, Duncan Vaughan, Peter H. Wood, Jessica Harris, Karen Hess, Haripriya Rangan, Ghislaine Lydon, Randy Sparks, Kent Mathewson, Marcus Rediker, Andrew Sluyter, Robert Keith Wayne, David Eltis, Jeffrey Pilcher, Tom Gillespie, Susanna Hecht, Dirk HilleRis-Lambers, Yongkang Xue, Robert Harms, Dorothea Bedigian, Lydia Pulsipher,

Cymone Fourshey, T. H. Culhane, Kevin Dawson, Starr Douglas, and José Almeida.

Rosa Acevedo Marin, Jacque Chase, António Crispim Veríssimo, and Fidelia Graand-Galon provided invaluable assistance with fieldwork in Brazil and Suriname. Scholars who generously helped with translations and matters of linguistic nuance are Rose Mary Allen, Quirino de Brito, Raul A. Fernandez, Sarah Fernández, Ana Paula Giorgi, Mark Moritz, and Alex van Stripiaan. Appreciation is also extended to John Moses, Gérard Second, H. Thoden van Velzen, and W. van Wetering. Special thanks to Chase Langford for his patience and thoughtful assistance in the preparation of the maps and images. His efforts proved indispensable to this project.

The book draws on some materials that appeared as published articles: "Rice and Memory in the Age of Enslavement: Atlantic Passages to Suriname," *Slavery and Abolition*, 26, no. 3 (2005): 325–47; "'With Grains in Her Hair': Rice History and Memory in Colonial Brazil," *Slavery and Abolition*, 25, no. 1 (2004): 1–27; "African Traditional Plant Knowledge in the Circum-Caribbean Region," *Journal of Ethnobiology*, 23, no. 2 (2003): 167–85. Appreciation is extended to the journals for permission to use these materials in revised form.

Finally, the anonymous reviewers of the proposal and manuscript cannot go unacknowledged. Their comments were uniformly sincere and encouraging. The metamorphosis from manuscript to book was guided by the sure and able hands of Jenny Wapner, Kate Warne, and Julie Van Pelt at the University of California Press. This book would not have been realized without their tireless efforts.

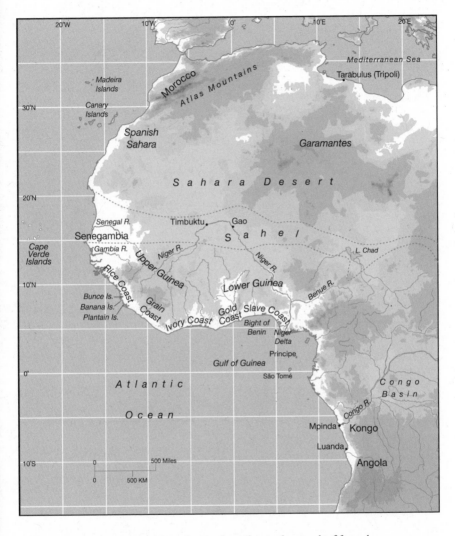

FIGURE 0.1.   Western Africa during the Atlantic slave trade, fifteenth–
nineteenth centuries.

FIGURE 0.2.   New World plantation societies, sixteenth–nineteenth centuries.

Introduction

The last half-century has seen a vigorous effort by scholars to unveil two partly obscured histories of the post-Columbian era, one of plants and animals and the other of peoples. . . . Yet at times it seems almost as if the histories of the peoples, at least in relation to the migration of flora and fauna, may have lagged behind—particularly if the peoples were not European.

SIDNEY W. MINTZ

THE POPULAR IMAGE OF AFRICA today is of a hungry continent, a continent chronically unable to feed itself, one that continually requires massive infusions of charity to keep its citizens from starvation. Yet this was not always so. In the thousands of years before the advent of recorded history, African peoples embarked upon the process of plant and animal domestication. As with their counterparts in tropical Asia and the Americas, Africans participated fully in the agricultural revolution that was taking place simultaneously around the world. By four thousand years ago, African food plants were on the move. They crossed the Indian Ocean and revolutionized the agricultural systems of semi-arid Asia. African crops were again being dispersed during the early modern period of European overseas expansion. They found a new footing in the New World tropics, where they contributed important dietary staples and commodities to plantation societies.

For the most part, the African crops did not make the transoceanic journeys on their own. They depended on people to move them from one continent to another and, at times, on specialized knowledge to cultivate, process, and prepare them as food. The story of African crops is little known and poorly understood. Many of the continent's dietary staples are incorrectly attributed to Asian origin. Others are dismissed as food crops of minor significance. When an important plant is actually acknowledged as an African domesticate, outsiders—ancient Asian mariners, Romans, Muslims, and Europeans—are credited with recognizing its useful properties and promoting its dispersal. The histories of African plant movements are much obscured by these longstanding assumptions and biases. This is especially the case with the introduction of African food plants to the Americas.

The historical importance of New World maize and manioc, introduced from Portuguese ships to West African societies, reinforces the modern perception of a continent perpetually suffering from food shortages. However, during the same period that Amerindian crops arrived in Africa, African dietary staples went to the Americas. Their early appearance in New World plantation societies warrants a reevaluation of African food systems. This new perspective draws attention to the transatlantic slave trade as the motive force behind the diffusion of African foodstaples. Europeans purchased food grown in Africa to provision slave ships. Slavers even recognized the value of stocking African staples as victuals, believing mortality rates declined if captives were fed the food to which they were accustomed. For the captains of slave ships, the utility of African crops ended when they disembarked their victims at the auction blocks of the Americas. There begins another tale, one that addresses the role of Africans in establishing their food crops in plantation societies of the Americas.

In the New World, enslaved Africans confronted a stark work regime with no assurance of sufficient food. But Africans were heirs to a body of knowledge that included tropical agriculture, animal husbandry, and the skills to recognize wild plants of food and medicinal value. They thus arrived with critical expertise that could improve their odds of survival. The first generations of Middle Passage survivors collectively accomplished something extraordinary. They adopted the subsistence staples of Amerindians and instigated the cultivation of familiar African foods. With this fusion of crop traditions, they confronted chronic hunger and diversified the often monotonous diets imposed by slaveholders. In so doing, slaves Africanized the food systems of plantation societies of the Americas. This book tells how they did so.

Throughout the Atlantic slave trade, when millions of laboring youth were forcibly removed from Africa's agricultural fields and carried off to work the plantations and mines of the Americas, Africa continued to generate food surpluses. Even when agricultural production was periodically disrupted by slave raiders in search of captives, European ship captains continued their reliance on food produced in Africa to provision their vessels. Over the course of three and a half centuries, these ships landed more than ten million Africans in the Americas.

Drawing attention to the historical role of African flora and fauna and the knowledge systems of its peoples, this book takes a stance that contrasts with commonly held, unexamined views about Africa. The species that ancient Africans domesticated evolved during a long historical process of adapting plants and animals to diverse environments. This knowledge was generated on a continent that is for the most part tropical. African species subsequently were well suited to play a significant role in ensuring food availability in other tropical regions of the world. Africa's botanical legacy traveled with its peoples and improved the sustenance of millions across vast intercontinental trading networks. These networks spanned thousands of years and moved sequentially from the Indian Ocean through the Mediterranean, eventually reaching the Atlantic shores of the Americas. The African diaspora was one of plants as well as peoples.

During the past fifty years, our understanding of the post-Columbian era has been enriched by efforts to uncover the hidden histories of the plants, animals, and peoples that were swept up in the making of the Atlantic world. A landmark among these efforts was Alfred W. Crosby's 1972 book *The Columbian Exchange: Biological and Cultural Consequences of 1492*.[1] This seminal work chronicles the revolutionary changes wrought by the intercontinental movements of plants and animals that followed the voyages of Christopher Columbus. European contact with the New World ushered in not only a period of exploration and conquest, but also a period of biological exchange in which peoples, plants, animals, and germs were carried across vast ocean networks to the Americas, Europe, and Africa. There, they radically transformed the ecosystems of four continents.

Some of the transfers of the Columbian Exchange were deliberate and are today broadly recognized: tomatoes, maize, potatoes, and tobacco from the New World to the Old World; breadfruit from Polynesia to the Caribbean; seasonings and condiments from the Spice Islands to Europe; wheat, oats, and grapes from the Mediterranean to the Americas. Other transfers were unintentional—vermin, insect pests, disease organisms, and weeds—but with impacts no less profound. Crosby subsequently made another important observation.[2] Not all species transfers were mediated by naturalists and agents of trade. Ordinary people also instigated the geographical dispersion of plants and animals across the globe. These were European emigrants in new lands—people who were not functioning as administrators, soldiers, or scientists. With the biota that accompanied them from Europe, settlers transformed the environments of Australia, New Zealand, and South Africa into familiar landscapes—in Crosby's words, into "Neo-Europes."

Thanks to Crosby's work and the wealth of scholarship that followed, we now know considerably more about the introduced plants and animals that transformed New World landscapes and about the significance of Amerindian food-staples to European settlement. The Columbian Exchange has become indispensable to our understanding of the biological and cultural history of the era. But largely unexamined are the African food plants and animals that appeared early in New World plantation societies. There is little written about the importance of African dietary staples in the Americas and even less attention paid to the ways they were introduced and established. Perceptions of a continent populated by hapless farmers and herders in need of European instruction are inaccurate and fail to do justice to Africa's deep botanical legacy. This book is written in part to redress this imbalance.

The nearly four hundred years since the landing at Jamestown, Virginia, of a boatload of enslaved Africans underscores the significance of the African presence to European settlement of the Americas. Until the second decade of the nineteenth century, Africans crossed the Atlantic in greater numbers than Europeans. Enslaved Africans and their descendants were central to the economic development of the New World. But their contributions involved far more than providing the muscle behind it. They brought critical skills and knowledge. Under conditions that today are scarcely imaginable, and seldom faced by any other immigrants to the Americas, slaves revitalized familiar foodways that were lost along with their freedom. On many different levels, the transatlantic slave trade represents one of the most important migrations in human history.

We live in a globalized and interconnected world, where the agricultural societies of just a century ago appear extremely remote from our own time. The tangled histories of the Americas, and the remarkable accomplishments of its own indigenous peoples in domesticating food crops, overshadow and complicate consideration of African contributions. Many of the cultivation methods, plant usage, and forms of food preparation developed by Amerindians survived their annihilation. Less appreciated is the role of enslaved Africans as the custodians of these knowledge systems to which they added their own farming acumen. The outcome of their efforts was the fusion of two tropical farming systems, one with roots in Africa and the other in the Americas. An experimental process that led initially to plant domestication in Africa did not end with the forced migration of Africans to plantation societies. Instead, it found new expression in the dooryard plots and food fields that enslaved Africans cultivated for subsistence.

The African components of intercontinental plant exchanges only come into focus by looking beyond the standard emphasis on commodities that dominates discussion of the burgeoning Atlantic world to a new emphasis on *subsistence,* which is to say, the everyday foods that sustained the human beings willingly and unwillingly involved in that era's economic expansion. In exposing the underlying importance of food to the realization of European hegemony, subsistence lays open the belly of the transatlantic slave trade. But it also draws attention to the ways that specific foods, no matter how humble in the eyes of outsiders, served to stave off starvation while also shaping the historical memory of Atlantic slavery's victims. To look at these plants is to engage the organization of the slave trade and plantation societies as corridors for the diffusion of African plants to the Americas; it is also to engage their symbolic meaning in shaping the identities of diasporic Africans. A new emphasis on subsistence recalls a time when nearly everything came from nature: food, medicines, cordage, housing, and clothing.

ONE

Food and the African Past

Anansi, the great spider of venerable memory, gathered all of the world's wisdom into a gourd. Seeking to safeguard this wisdom for future generations, he hung the gourd high in a tree. But he failed. The gourd fell and broke, and wisdom was scattered far and wide.

Akan proverb

Unumbotte made a human being. Its name was Man. Unumbotte next made an antelope, named Antelope. Unumbotte made a snake, named Snake. At the time these three were made there were no trees but one, a palm. Nor had the earth been pounded smooth. All three were sitting on the rough ground, and Unumbotte said to them: "The earth has not yet been pounded. You must pound the ground smooth where you are sitting." Unumbotte gave them seeds of all kinds, and said: "Go plant these."

Bassari legend

OUR AWARENESS OF AFRICA BEGINS as the place where our hominid ancestors evolved. But it remains a "Dark Continent" in terms of broader understanding of what African peoples accomplished in the millennia preceding the transatlantic slave trade, when the continent's history again comes into focus for modern audiences. Unexamined views of Africa carry the presumption of a continent on the sidelines of world history, where little occurred until the Atlantic slave trade swept away millions of its people. In the New World, so these views maintain, Europeans taught their unskilled bondsmen to plant crops and tend animals. Nevertheless, one of the remarkable achievements of Africans over the past ten thousand years was the independent domestication of plants and animals for food. These African species journeyed to Asia in the second and first millennia B.C.E. and, later, profoundly shaped the food systems of plantation societies of the Americas.[1]

The African continent harbors more than two thousand native grains, roots and tubers, fruits, vegetables, legumes, and oil crops. Its plant resources also provide stimulants, medicines, materials for religious practices, and fodder for livestock. Long before Muslim caravans and Portuguese caravels reached the African continent, its peoples initiated the process of plant and animal domestication. Africans have contributed more than one hundred species to global food supplies.[2] The plants they gave the world include pearl (bulrush) millet, sorghum, coffee, watermelon, black-eyed pea, okra, palm oil, the kola nut, tamarind, hibiscus, and a species of rice. Widely known consumer products— Coca-Cola, Palmolive soap, Worcestershire sauce, Red Zinger tea, Snapple and most soft drinks—rely in part on plants domesticated in Africa.[3] African contributions to global plant history, however, are largely unacknowledged and seldom appreciated. In the popular image, Africa is a place of hunger and starvation, a continent long kept alive by food imported from other parts of the world.

But this modern perception belies a very different history. In ancient times, African cereals transformed the food systems of semiarid India by providing grain and legumes suitable for cultivation. Domestication of animals and plants in Africa thousands of years ago instigated a continuing process of indigenous experimentation and innovation, a process that incorporated species later introduced from other continents. From these immigrant species, Africans developed new cultivars and breeds that strengthened the capacity of food systems to provide daily subsistence. By the time Europeans visited the west coast of Africa in the fifteenth century, they found a land whose bounty provoked admiring commentary. In the words of the seventeenth-century Luso-African trader Lemos Coelho: "The blacks have many foodstuffs such as [guinea] hens, husked rice (all high-quality and cheap), plenty of milk, and excellent fat (*manteiga*, 'butter'). . . . This is because the whole kingdom of Nhani is full of villages of Fulos [Fula], who have these foodstuffs in abundance. A cow costs only a *pataca* or its equivalent. . . . Thus everything necessary for human existence is found in this land in great plenty and sumptuousness."[4]

Even as the transatlantic slave trade removed millions of people from the continent, African societies produced food surpluses. And in the Americas, enslaved Africans continued their innovating processes. They nurtured Africa's principal dietary staples in their food fields and adopted Amerindian crops beneficial to their survival. Livestock-raising peoples such as the Fula brought animal-husbandry skills that contributed critically to New World ranching traditions.

Human beings evolved in Africa. Fully modern humans, *Homo sapiens sapiens*—people like us with full syntactical language—emerged from the African ancestral line of all living humans between 100,000 and 60,000 years ago.[5] This evolutionary development probably took place in the eastern parts of Africa. With their new capacity for language, these first true humans soon spread across the continent. Later, between 60,000 and 50,000 years ago, a small group of these fully human hunter-gatherers left Africa. They expanded outward from northeast Africa following one, possibly two routes: with watercraft along the Indian Ocean shores of southern Asia and by foot across the Sinai Peninsula into the Middle East.[6] The descendants of these African emigrants eventually settled most of the habitable regions of the world, giving rise to the human populations that we recognize today.

The DNA of human cells preserves a biological record of this remarkable global journey. Two different genetic signatures—one from the DNA of cellular mitochondria (mtDNA) and the other from the Y-chromosome that confers maleness—offer compelling scientific evidence for humanity's "out of Africa" origins. MtDNA is exclusively inherited from our mothers, so mutations pass intact from one generation to the next through the female line. The extent of change or variation on a strand of mtDNA may be quantified by chemical analysis. Through this technique, mtDNA provides geneticists a tool for measuring the evolutionary distance between the original African population of human beings who evolved on the continent and those descended from the small group who formed its primordial diaspora. Scientific studies of the human genome show that a nearly full diversity of all mtDNA variation occurs in the populations of Africa. MtDNA lineages found in humans outside Africa belong to one subset of that diversity. A similar pattern characterizes the Y-chromosome, which men inherit directly from their fathers. Y-chromosome lineages found outside Africa comprise one subset among all the Y-chromosome lineages found among Africans. The subsets of mtDNA and Y-chromosome lineages found in humans outside the continent are most typical of northeastern African populations. This supports the conclusion that human populations of the rest of the world originated through the migration of peoples out of that part of the continent. Africa is indelibly imprinted in the genetics of all human beings.

One of the greatest achievements of human beings was the domestication of plants and animals. Just as the earliest *Homo sapiens sapiens* of 90,000 to 60,000 years ago coped with different environments as they spread across Africa, so did their descendants who fanned out across the rest of the globe after 60,000 years ago. The African emigrants encountered diverse new environments in which they discovered edible plants and new animal food sources. On grasslands, river floodplains, highlands, marshes and coastal estuaries, they hunted animals and birds, took fish from bodies of water, gathered wild plants, and uprooted edible tubers. In distinctive environmental settings, humanity began the long process of manipulating species for their food, medicines, and spiritual needs. Some ten thousand years ago the process of plant and animal domestication was simultaneously underway in several parts of the world, including Africa.

Africa is the world's second-largest continental landmass after Asia. Three times the size of Europe and larger than North America, Africa is nearly an island. Just a sliver of land connects the continent to the Sinai Peninsula and the Near East. It is a sprawling continent, extending across 72 degrees of latitude. Most of its landmass falls within the tropics, between the defining lines known as the Tropics of Cancer and Capricorn. The equator cuts the continent into two unequal halves: from Tunisia lying 37 degrees latitude north to Cape Town at 34 degrees south, the distance traversed is equivalent to that between New York and Hawaii. Within these vast geographical extremes, life, settlement, and livelihood strategies have been adapted to heterogeneous landscapes.

Africa is a mostly tropical continent, but it is also a savanna continent. The most extensive areas of savanna in the world are presently found in Africa south of the Sahara Desert. These grasslands cover areas of low topographic relief that include diverse types of environments. Their vegetative profile varies with rainfall, soil type, and human activities. Some savannas are well watered, others not. This gives them their distinctive appearance, which ranges from thickly wooded grasslands to nearly treeless plains. Many of Africa's principal food crops were domesticated in these variegated landscapes.

The savanna environment is also in part the result of human agency. Africa's savannas have long been manipulated by fire, both natural and human-made. In the thousands of years prior to the onset of the domestication of wild species, human beings used fire to drive game animals in desired directions. Burning also encouraged the growth of nutritious grass shoots that became forage for wild animals. When ancient peoples discovered this, they exploited the practice to improve the survival of the game they hunted. Today no other environment on the face of the earth supports animals of such spectacular size or herds of such immense numbers. Africa is often called the living Pleistocene precisely because of the game animals that thrive on its savannas.

Research from a number of disciplines suggests that Africa's pathway to food production may have been unique among world regions where the domestica-

FIGURE 1.1.   Cattle herd from Tassili Jabberen rock art, ca. 4000 B.C.E.
SOURCE: Lhote, *Search for the Tassili Frescoes,* pl. 27.

tion of plants and animals took place. Archaeological, genetic, botanical, and linguistic studies indicate that the earliest African food producers were likely mobile herders rather than sedentary plant gatherers. In contrast with other regions of the globe, the domestication of animals in Africa apparently preceded that of plants.[7]

Africans began the process of plant and animal domestication around 10,500 B.P. From the last glacial maximum some 20,000 years ago to the close of the ice ages, the Sahara was hyperarid, even more so than it is today. After 10,500 B.P. the Sahara's climate became wetter. In the eastern Sahara this transformation was both abrupt and dramatic. For three thousand years, from 8500 to 5300 B.C.E., summer rains from the Atlantic airflow system periodically

reached the region.[8] In the Nubian Desert of southern Egypt, a large basin known as the Nabta Playa began filling with water. The unpredictable rains and frequent droughts during this climatic phase made the emergent lake an attraction for animals and human beings. The earliest settlements at Nabta, some 10,500 to 9,300 years old, suggest the inhabitants were cattle-keeping people.[9]

The oscillating wet and dry periods that marked the early Holocene may have encouraged hunter-gatherers to domesticate cattle as a food source.[10] The uncertainties of environmental change and the need for more dependable food supplies perhaps prompted ancient Africans to embark upon the domestication of the wild cattle species, *Bos primigenius africanus*. The oldest undisputed remains of domesticated cattle in the eastern Sahara date to 8,000 years ago. However, the Nabta evidence suggests that domestication may have taken place between 10,500 and 9,000 B.P., more than a millennium earlier. Recent data from archaeology, historical linguistics, and DNA analysis now support a claim for an independent domestication of humpless longhorn cattle in the eastern Sahara at an early date.[11]

Cattle tending spread westward across Saharan grasslands and south to the Red Sea Hills.[12] Rock art of the western Sahara indicates that pastoralists diffused across the savannas of the Sahara between 8,000 and 6,000 B.P. (figure 1.1).[13] Around 5300 B.C.E., a decisive climate shift in the eastern Sahara promoted a gradual desiccation. The original area of cattle domestication in turn grew less hospitable to herders. The cessation of summer rains once again brought hyperaridity to the region. Pastoralists responded by moving their cattle herds to wetter savanna environments. The return of desert conditions in Egypt about 3500 B.C.E. coincided with the initial period of the pharaonic civilization, centered on the Nile River floodplain. By then the entire Saharan region was trending toward decreased precipitation.[14]

As aridity once again returned to the Sahara, lakes and water holes on the grasslands that had made the region attractive to game animals, hunter-gatherers, herders, and fishing peoples dried up. Eventually, rainfall declines shaped the Sahara into the arid landscape that it is today. Domesticated livestock provided herders a reliable subsistence strategy, as people and animals followed the patchy retreat of savannas to the verdant grasslands to the south and southwest, an area now known as the Sahel. Herding dramatically reduced the risk of hunger because the mobile food supply could be relocated to take advantage of local differences in forage and water availability.[15] Human populations came to depend on cattle for the milk and meat their animals produced. The Sahara's earliest food-producing communities thus were organized around the practice of herding.

Animal husbandry did not stop with the taming of wild cattle. Ancient herders continued the process of selecting and breeding animals with desirable traits. One breed with long bulbous horns, known as Kuri or Buduma cattle, developed around Lake Chad. It consumes aquatic plants for food and spends several hours each day immersed in water. The breed's long horns act as a flotation device. Herders developed another indigenous breed in the humid woodlands of Guinea's Futa Jallon plateau: the dwarf humpless cattle known as *n'dama,* whose outstanding virtue is resistance to the bovine sleeping sickness (trypanosomiasis) transmitted by the tsetse fly. The *n'dama* breed allows cattle-keeping in fly-infested areas that would prove lethal to animals without this special trait.[16] Ancient herders bred other types for aesthetic features. The continent's indigenous breeds are distinguished by slender, sinewy limbs and often by their horns, which are long, inward curving, and crescent- or lyre-shaped (figures 1.2, 1.3).[17]

FIGURE 1.2.   Indigenous African *n'dama* cattle, northern Sierra Leone, 1967.
Photo courtesy of David P. Gamble.

Cattle were not the only animals that ancient Africans domesticated in arid
environments. Between 7,000 and 5,000 B.P. Africans developed the donkey
(Equus asinus) as a transport animal. The animal enabled early pastoralists to
respond to growing aridity through frequent relocations between camps and
oases. The donkey's capacity to carry wood, water, and people over short dis-
tances, made it an efficient transportation system. It was likely domesticated
from the indigenous wild ass *(E. africanus),* which occupied the savanna steppes
that then stretched from the Horn of Africa westward to the Atlas Mountains
(now part of the Sahara).[18] Eventually, the donkey spread from the African continent to the Near East and Mediterranean world. Another savanna food animal that Africans domesticated was the guinea fowl *(Numida meleagris),* the
continent's indigenous poultry species.[19]

The experimental process that led Africans to domesticate cattle and develop
breeds adapted to specific ecological and aesthetic contexts was applied to other
livestock introduced to the continent. Sheep and goats arrived in the Saharan
regions by 8,500 years ago, camels in the first millennium B.C.E., and the hump-

FIGURE 1.3.   Indigenous African *n'dama* cattle, the Gambia, 1963. Photo courtesy of David P. Gamble.

less Indian zebu cattle about 2,000 years B.P. Each of these domesticated species enabled African pastoralists to diversify their herd composition and breeding stock. With drought a recurrent threat in these regions, ownership of many different types of livestock reduced the chances of losing everything, diffusing the risk among different breeds and species. Herd diversification moreover improved the forage efficiency of savannas because each species grazed different parts of grasses or consumed plants unpalatable to the others. The introduction of these food animals involved African herders in a continuous process of experimentation as they adapted the species to their needs. They crossed the indigenous longhorn cattle with the zebu to develop a hardy drought-tolerant breed known today in Uganda as ankole. Herders also bred a type of woolless thin-tailed sheep that was particularly adapted to the continent's hot arid savannas but that was also able to survive in humid tropical areas. This so-called hair sheep was raised for meat rather than fiber. A distinct fat-tailed and also nonwoolly sheep, adapted to a variety of savanna and mountain environments, became the predominant breed across the eastern side of the continent.[20]

As livestock keeping spread throughout the sub-Saharan region, so did the indigenous grasses that provided forage. African pasture grasses are adapted to tropical growing conditions. Some flourish in well-drained soils, others in humid bottomlands, still others in sandy soils of arid landscapes. When the Portuguese and Spanish arrived along the western coast of Africa in the fifteenth century, these native forage and fodder grasses facilitated animal husbandry in many diverse tropical settings. Spurred by the needs of burgeoning colonial enterprises, European ships carried the continent's food animals and pasture grasses as live provision and breeding stock to the New World tropics.

AGRICULTURAL DOMESTICATION IN AFRICA

Domestication is the selection process by which human populations adapt plants and animals to their need for a predictable food supply. It is a symbiotic relationship, as domestication allows species to spread beyond their original geographical boundaries in exchange for their use as food by their benefactors. Plant and animal domestication began after the last ice age and developed independently in several world regions, including Africa. Linguistic and archaeological evidence suggests that people in two parts of Africa—the West African savannas and the southern eastern Sahara—began cultivating plants as early or nearly as early as people in southern East Asia, northern China, the Near East, interior New Guinea, and Mesoamerica. However, ancient Africans for a long time allowed their cultivated crops to interbreed with the wild forms, thus slowing the appearance of fully domesticated varieties. The presence of tropical African domesticates such as melons (already in Egypt by the third millennium B.C.E.) and finger millet and pearl millet (in India by the second millennium B.C.E.) requires, in any case, that African agriculture developed well before these dates.

The process of agricultural domestication in Africa, as elsewhere, involved the selection of wild plants that displayed fundamental features esteemed by human populations. These traits included larger size, higher yield, controlled ripening, palatability, and ease of processing. Importantly, the morphological changes that accompanied the transformation of a wild plant to its domesticated progeny took a very long time to accomplish. Domestication was a slow and laborious process, one that may have in the African case taken two or three millennia to realize.[21] A Saharan rock painting, dating to circa 4000–1500 B.C.E. (plate 1), depicts women apparently gathering grain. Considering the extremely arid conditions that prevailed in the Sahara by that period, they may be collecting wild grain that sprouted after a rare rain, since farming by then would have been risky if not impossible.

Current evidence from archaeology, historical linguistics, and botany suggests that plant domestication was initiated in Africa during the millennia before 4,000 B.P. Just how long before is an issue still to be resolved. In northeastern Africa, archaeological research suggests that the cultivation of sorghum was underway in Sudan by 7,000–6,000 B.P.[22] There is also indirect evidence for the domestication of cotton by 5000 B.C.E., with spindle whorls recovered from archaeological sites near Khartoum.[23] Indirect evidence of African plant domestication has been found in the archaeological record of the West African rainforest. The spread of polished stone axes through the rainforest, as tools for clearing the forest for the cultivation of yams, dates to 5000–4000 B.C.E.[24] Linguistic evidence from both regions implies that food plants were cultivated as much as 2,000 years before these indirect archaeological indicators. But we still lack in most areas sufficient archaeological research to substantiate the cultivation of plants before 5000 B.C.E. The cumulative weight of this evidence from archaeobotany and linguistics nonetheless suggests the likelihood that plant cultivation is far more ancient on the African continent than has been supposed.[25]

In Africa, as elsewhere, agricultural domestication resulted from experimentation with wild plants that people had gathered for millennia. All the foodstaples of the world are based on the achievements of our collective forebears, who identified in widely divergent environmental settings edible foods and plants with desirable properties. In Africa, ancient plant gatherers initiated the domestication process with the wild grasses, fruits, nuts, greens, legumes, and underground tubers that they had long collected for food. From this botanical assemblage, they created the plants' domesticated progeny. Despite the turn toward agriculture, the foraging of wild botanical resources continues to provide contemporary populations important supplemental sources of food and medicine, especially in areas of unpredictable rainfall. On the savannas of West Africa alone, more than two hundred plants are still foraged for their edible fruits, seeds, and leaves.[26]

Most of the plants ancient Africans domesticated are adapted to a wide range of tropical farming conditions. Several are drought-tolerant; others mature rapidly or possess specific traits that promote food production in diverse ecological settings. Pearl millet, for example, produces a harvest in just a few months; sorghum grows more slowly and is able to mature in the dry season, relying on the residual moisture held by the soil. Although the yam is considered a tropical forest crop, it originated on the western savannas of the continent. Thought to be one of the first food crops Africans domesticated, the yam was subsequently developed as an important tuber of the humid tropical forests of Nigeria and Cameroon.[27]

Ancient Africans fully met the challenges of a dramatically changing Holo-

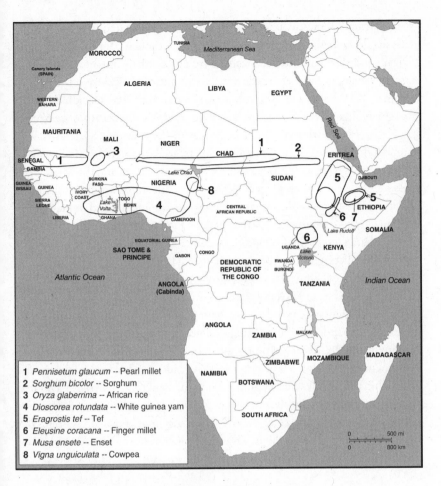

FIGURE 1.4. Areas of domestication of selected African crops.

SOURCE: Adapted from Harlan, "Agricultural Origins," 471.

cene landscape by adapting their subsistence strategies. Agricultural origins in Africa took place in lowlands and hilly regions, in various types of grasslands, wetlands, and forests, and under a variety of ecological conditions. On the African savannas, plant domestication occurred over a broad region rather than in a specific geographic center of diversity. This noncentric pattern of domestication contrasts with most other regions of Africa and the world, where agriculture spread from a single geographic locus. Figure 1.4 depicts the geographical contours of these patterns of African plant domestication for some of the continent's principal foodstaples.[28]

Agricultural domestication was a dynamic and variegated process. Ancient Africans developed different subspecies of each domesticate for specific ecological conditions and human needs. Sorghum, perhaps the most ancient of African cereals, is particularly illustrative: from its wild origins in the African savanna, there developed more than two dozen cultivated species. Some are drought-tolerant, others are adapted to river floodplains, where they are planted in soils exposed during the seasonal retreat of flood waters. Sorghum's environmental versatility is matched by its many uses. Some types yield a sweet, molasses-like syrup; the grain may be malted to brew beer, and the whole plant makes excellent livestock silage.[29] Pearl millet—the most drought-tolerant of all cereals—reflects a similar adaptability to environmental and human needs. African farmers developed types that enabled food production right to the edge of the Sahara Desert. Such adaptations contributed to shaping the noncentric pattern that distinguishes African agricultural beginnings in the savannas.

As the Holocene climate changed, farming gave African peoples a second way to improve food security. Agriculture contributed a crucial livelihood strategy to the diversified systems of food procurement that human populations increasingly demanded. By supporting a range of food-procurement strategies— hunting-gathering, herding, fishing, and farming—the savannas enabled the development of specialist ethnic groups.[30] The wide variance in rainfall between and within years made such adaptations, and the food exchanges they facilitated, vital to the collective survival of the continent's peoples. Plant domestication added yet another crucial component to the indigenous knowledge systems that had developed on the continent. Thousands of years later, these African faunal and botanical species would play a role in the history of Atlantic slavery.

ECOLOGICAL COMPLEXES
OF AFRICAN PLANT DOMESTICATION

Plant domestication in Africa represents a significant contribution to world agriculture. Africans added three important cereals, a half-dozen root crops, five oil-producing plants, more than a dozen leafy vegetables and greens, about a half-dozen forage crops, a variety of beans, nuts, and fruits, in addition to the versatile gourd (used as a container, for musical instruments, and fishing floats).[31] Many of these crops are today vital to the sustenance of millions living in tropical areas around the world.

Agricultural domestication occurred in three distinct ecological complexes of sub-Saharan Africa: the highland to lowland gradient of Ethiopia; along the savannas that stretch nearly unbroken across the continent from Sudan to Mauritania; and the forest-savanna ecotone of present-day Cameroon and Nigeria in west-central Africa. Figure 1.5 lists the principal African food domesticates from each of these distinctive environmental settings.

Plant domestication unfolded in the highlands of Ethiopia and Uganda and in the southern Ethiopian lowlands. Ethiopia is the birthplace of the world's premier coffee *(Coffea arabica)*. Other important food crops domesticated in the region include finger millet *(Eleusine coracana)*, tef (the cereal staple of Ethiopian cuisine), the lesser-known enset (a root crop that resembles the banana), the lablab or hyacinth bean (also known as bonavist), and the castor bean oil plant.[32] Finger millet, the lablab bean, and the castor plant diffused widely across Africa and the ancient world. Castor bean was esteemed for its dual use as a lamp oil and medicinal.[33]

The savannas north of the equator hosted the domestication of many African foodstaples that have become significant outside the continent. This ecological region, of such significance in the story of African cattle domestication and the ancient migration of pastoralists from the eastern Sahara, is central to the development of many of the world's most important tropical cereal and pasture grasses. Sorghum *(Sorghum bicolor),* for example, today ranks fifth in world cereal grain production. It produces a larger grain than pearl millet. The sorghum plant physically resembles maize in its vegetative stage, but instead of producing an ear its stalks are crowned by a plumed head of seed grains. It is typically made into porridges of varying thickness and brewed into beer.

The archaeobotanical evidence sets the domestication stage in the history of pearl millet *(Pennisetum glaucum)* cultivation no later than 3,500 years ago.[34] Thriving in arid climates where no other cereal grows, the plant produces exposed grains along a stalk that resembles a bulrush or cattail. Pearl millet is the most drought-resistant of the world's principal food grains; no other cereal produces such reliable yields under such challenging conditions of withering heat and minimal rainfall. For precisely these reasons, it is the dietary staple of millions living in arid tropical regions. In Africa the cereal's grains are ground as flour or cooked as a whole grain. Pearl millet is prepared in a variety of ways, as unleavened bread, porridge, fermented snacks, and steam-cooked couscous. In the mountainous interior of Niger, pastoralists long ago learned to prepare it for desert journeys without the need for additional cooking. The grain is roasted and mixed with dried dates and goat cheese. The millet-based mixture provides a nutritious dietary staple on long caravans across the Sahara and obviates the need to carry firewood and water for meal preparation.[35]

An even smaller grained type of millet is fonio *(Digitaria exilis, D. iburua),* or "hungry rice."[36] More drought-tolerant than pearl millet, this African grain

FIGURE 1.5. Food crops of African origin.

SAVANNA COMPLEX

Adansonia digitata L.	Baobab
Brachiaria deflexa (Schumach.) C. E. Hubb. ex Robyns	Guinea millet
Ceratotheca sesamoides Endl.	False sesame: leaves and seeds
Citrullus lanatus (Thunb.) Matsum & Nakai	Watermelon
Corchorus olitorius L.	Jute mallow/bush okra
Cucumis melo L.	Muskmelon
Digitaria decumbens Stent	Pangola grass
Digitaria exilis (Kippist) Stapf	Fonio/"hungry rice"
Digitaria iburua Stapf	Black fonio
Hibiscus cannabinus L.	Kenaf
Hibiscus sabdariffa L.	Roselle/hibiscus/*bissap*
Lagenaria siceraria (Molina) Standl.	Bottleneck gourd
Oryza glaberrima Steud.	African rice
Parkia biglobosa (Jacq.) R. Br. ex G. Don	Locust bean
Pennisetum glaucum (L.) R. Br.	Bulrush or pearl millet
Polygala butyracea Heckel	Black beniseed
Sesamum alatum Thonn.	Sesame (leaves)
Sesamum radiatum Schumach. & Thonn.	Beniseed
Solanum aethiopicum L.	African eggplant/garden egg/guinea squash
Solanum incanum L.	Bitter tomato
Solanum macrocarpon L.	Nightshade
Sorghum bicolor (Linn.) Moench	Sorghum/guinea corn
Vigna subterranea (L.) Verdc.	Bambara groundnut/*Voandzeia*
Vitellaria paradoxa C. F. Gaertn.	*Karité* or shea nut tree
Xylopia aethiopica (Dunal) A. Rich	Guinea pepper

WEST AFRICAN SAVANNA-FOREST COMPLEX

Aframomum melegueta K. Schum.	Melegueta pepper
Amaranthus spp.	Vegetable amaranth/African spinach/*bledo*/*callalou*

SOURCES: Harlan, *Crops and Man,* 71–72; MacNeish, *Origins of Agriculture,* 298–318; Vaughan and Geissler, *New Oxford Book of Food Plants,* 10, 26, 38, 128, 174; Marshall and Hildebrand, "Cattle before Crops," 123–24; Tropicos, www.tropicos.org; Aluka, www.aluka.org.

Blighia sapida K. D. Koenig	Ackee/akee apple
Cajanus cajan (L.) Millsp.	Pigeon pea/Congo pea/Angola pea/*guandul*
Coffea robusta Linden	Coffee (robusta)
Cola acuminata (P. Beauv.) Schott & Endl.	Kola nut
Cola nitida (Vent.) Schott & Endl.	Kola nut
Cucumeropsis edulis (Hook. f.) Cogn.	*Egusi*
Dioscorea bulbifera L.	Air potato yam
Dioscorea cayenensis Lam	Yellow guinea yam
Dioscorea dumetorum (Kunth) Pax	Three-leaved or bitter yam
Dioscorea rotundata Poir.	African yam
Elaeis guineensis Jacq.	Oil palm/*dendê*
Gossypium herbaceum L.	Cotton
Hibiscus esculentus L.	Okra/gumbo
Kerstingiella geocarpa Harms	Kersting's groundnut/Hausa ground nut/geocarpa bean
Momordica charantia L.	African cucumber/bitter melon/balsam pear
Piper guineense Schumach. & Thonn.	Piper seed
Plectranthus esculentus N. E. Br.	Dazo/fingerling potato
Solenostemon rotundifolius (Poir.) J. K. Morton	Hausa potato/piasa
Sphenostylis stenocarpa (Hochst. ex A. Rich.) Harms	African yam bean
Tamarindus indica L.	Tamarind
Telfairia occidentalis Hook. f.	Fluted pumpkin
Vigna unguiculata (L.) Walp.	Black-eyed pea/cowpea/calavance

ETHIOPIA/EAST AFRICAN HIGHLANDS COMPLEX

Avena abyssinica Hochst.	Ethiopian oats
Catha edulis Forssk.	Chat
Coccinia abyssinica (W. & A.) Cogn.	Anchote
Coffea arabica L.	Coffee (arabica)
Eleusine coracana (Linn.) Gaertn.	Finger millet
Ensete ventricosum (Welw.) Cheesman	Enset
Eragrostis tef (Zucc.) Trotter	Tef
Guizotia abyssinica (L.f.) Cass.	Niger seed, noog
Lablab purpureus Sweet	Lablab/bonavist/hyacinth bean
Panicum maximum Jacq.	Guinea grass
Ricinus communis L.	Castor bean

thrives on poor savanna soils. It was of such importance in Sahelian food systems that Ibn Battūta described it on his visit from southern Morocco to Mali in 1352, where it was made into couscous. While fonio can be harvested in just six to eight weeks, its very small grains make removal of the husk quite difficult, which has limited the adoption of this otherwise nutritious grain outside Africa.[37]

Another significant cereal domesticated in the savanna landscape of West Africa is African rice *(Oryza glaberrima)*. It is distinguished by a reddish hull that encloses the grain. The domestication of rice took place later in Africa than in Asia, but it unequivocally occurred in West Africa prior to the introduction of Asian rice *(O. sativa)* to the African continent. African rice was present at the beginning of human occupation at Jenné Jeno along the interior delta of the Niger River in Mali some two thousand years ago.[38] Archaeobotanical data from this and other sites in the region increasingly support a date no later than 3,500 B.P. for the domestication of African rice.[39] The linguistic evidence suggests that the cultivation of this crop began even earlier, as much as 5,000 years ago.[40] Secondary centers of rice domestication are believed to have developed north and south of the Gambia River and in the Guinean Highlands. This has contributed to the broad extension of West Africa's indigenous rice region southward from the Senegal River to the Ivory Coast and inland all the way to Lake Chad.[41]

The harsh environment of the drier African savannas supported the ancient domestication of the African eggplant *(Solanum aethiopicum)*, prized for its fruit and leaves.[42] The region is also home to several cucurbits—plants that include melons, gourds, squashes, and cucumbers. Many of these species were esteemed in ancient times not so much for their fruit, but for their edible roasted seeds. They remain important as thickeners and flavoring agents in many West African dishes, where they are known collectively as *egusi* melons.[43] Africa is the continent where the ancestor of the watermelon was domesticated. Its prototype was originally a bitter melon that was grown on arid savannas for its edible seeds and as a storable form of moisture.[44] Africa is also believed to be the original home of the cantaloupe or muskmelon *(Cucumis melo)*. The bottleneck gourd *(Lagenaria siceraria)*, another cucurbit species considered of African origin, has been known since ancient times for its edible seeds, utility as a container and ladle, and use as a percussive or stringed musical instrument. The gourd's fundamental cultural importance is likely the reason it often serves as a symbol and metaphor in oral and written traditions. As the chapter's epigraph indicates, the gourd is the vessel of wisdom in many African legends. In plantation societies of the American South, the African gourd symbolized freedom, as in the African American song, "Follow the Drinking Gourd"—in reference to the celestial Big Dipper, whose stars guided passengers of the Underground Railroad to the north and freedom.[45]

Africa hosts wild and semicultivated species of sesame, which are grown for their leaves as much as their seeds. The sesame crop of international commerce (*Sesamum indicum*) was domesticated several millennia ago on the Indian sub-continent.[46] Africans domesticated the protein-rich Bambara groundnut (*Vigna subterranea*), which grows underground like peanuts and is of exceptional nutritional quality. Another savanna plant widely known beyond Africa is hibiscus (*Hibiscus sabdariffa*). It produces a refreshing beverage with a cranberry-like flavor and color. Known as *bissap* in Senegal, where it is the national drink, throughout Latin America it is called *flor de Jamaica*. Hibiscus is an important ingredient of many commercial drinks and herbal teas. Finally, the savanna hosted the domestication of the most important medicinal plant found in African-descended regions of the Americas, the bitter melon (*Momordica charantia*). Sometimes referred to as Chinese bitter melon, the African plant's medicinal values since ancient times have also been appreciated in Asia for a wide variety of cures.[47]

The African savanna supports a variety of trees that since antiquity have provided humans with useful by-products. Farmers have long protected these species as sources of food, oil, resins, cordage, and medicines. Chief among these savanna species are the locust bean (*Parkia biglobosa*), baobab (*Adansonia digitata*), gum arabic (*Acacia senegal, A. seyal*), and the shea nut tree (*Vitellaria paradoxa*). Gum arabic, today important as a food stabilizer and an ingredient of most of the world's soft drinks, was among the first African food products known in medieval Europe.[48] It was used as a medicinal and as a component of inks and paints. The shea nut tree has provided a cooking oil and moisturizer to savanna populations for at least a thousand years. Today it is promoted by the global cosmetics industry in natural skin care products.

The grassland steppes—so central to the story of African cattle and the ancient pastoralists who tended them—are the original home of many of the world's most important tropical pasture grasses. These include several with suggestive common names: Angola or Pará grass (*Brachiaria mutica*), guinea grass (*Panicum maximum*), Bermuda grass (*Cynodon dactylon*), molasses grass (*Melinis minutiflora*), and thatching grass (*Hyparrhenia rufa*). Besides their use as fodder, several of these grasses have applications in traditional medicines.[49]

The savanna-to-forest transitional zone found in Cameroon and Nigeria provided the third crucial setting for the domestication of African plants. Located between the southernmost extension of cattle ranching and evergreen forest, this ecologically diverse woodland savanna is believed to support the richest diversity of flora and avian species on the continent. The West African agricultural tradition likely began with the yam, where natural clearings allowed the plant to thrive. Among this region's indigenous plants are many that provide dietary protein, such as the cowpea, pigeon pea, and one lesser-known legume—Hausa, or Kersting's groundnut, grown for its edible underground nut.[50] In Africa, these nutritive plants provide food for both people and animals. The cowpea is a cousin of the "Asian" long bean, which is an important ingredient in Far Eastern cuisine.[51] Known in North America as the black-eyed pea, the cowpea is frequently intercropped with sorghum. Sorghum's tolerance of infertile soils perfectly complements cultivation of this legume, whose nitrogen-fixing properties restore depleted nutrients.

The forest-savanna ecotone harbored other plants that have figured prominently in food traditions outside the continent. These include okra, the tamarind, the oil palm (the signature ingredient of Brazilian Bahian cooking and used also to make Palmolive soap), and the yellow yam *(Dioscorea cayenensis)*. Initially domesticating them from a wild savanna species, African farmers developed several cultivars of yam *(D. cayenensis, D. rotundata, D. dumetorum, D. bulbifera)* for cultivation in the wet tropics. Planted in aerated earthen mounds, the yam produces with little effort large tubers that can be stored for later consumption. It has long served as a principal dietary staple of humid tropical Africa. The tuber is also widely recognized for its medicinal properties. Its leaves are chewed to relieve gastric distress; the root provides steroids with anti-inflammatory properties that reduce cholesterol levels, swelling from arthritis and rheumatism, and fungal growth on human skin. Its value as a food crop and medicinal has long been vested with symbolic meaning through celebrations known as yam harvest festivals. Europeans first encountered the yam in West Africa.[52]

Many important stimulants, medicinals, and spices also originated in the deciduous and evergreen forests of western Africa. These include several plants whose African origin is seldom recognized. Two of the most widely consumed beverages in the world today are based wholly, or in part, on plants ancient Africans domesticated: the kola nut (a principal ingredient of Coca-Cola) and coffee. Other plants of African origin include melegueta pepper and the ackee apple, which along with salt fish is the national dish of Jamaica.[53]

AFRICANS AS TROPICAL FARMERS AND HERDERS

The cultivation of food in the wet tropics relies upon markedly different principles than that of temperate farming systems. What appears as luxuriant natural vegetation often masks soils of limited fertility. Most tropical plant nutrients are instead locked up in the vegetative mantle. Once an area is cleared for agriculture, soil fertility undergoes rapid decline. In the era prior to chemical fertilizers, fertility was restored by burning secondary growth for ash, adding organic matter such as animal manure and crop residues, planting beans and leguminous trees that make atmospheric nitrogen available to crops, and by leaving the land periodically fallow. The high year-round temperatures of the tropics moreover contribute to the proliferation of insect pests. Farmers do not benefit from the seasonal climate swings of temperate zones, where colder weather reduces insect numbers. Heavy tropical rainfall threatens the erosion of valuable nutrients when soils are cleared for agriculture. Considerable care must be taken to prevent the percolation of valuable plant nutrients down the soil horizon, which would put them beyond the reach of crop roots.

To minimize these agricultural constraints, farmers across the Old and New World tropics domesticated a diverse array of plants to realize their food needs. Over millennia of trial and error, they devised ingenious ways to feed themselves without destroying the soil that supports crop growth. Then, as now, they often planted in multicropping systems—different species and cultivars grown together on the same plot—intercropping the cultivation of seed plants, tubers, and legumes with valuable fruit-, nut-, and oil-bearing trees. This crop diversity diminishes the ability of insects to destroy an entire food system. It encourages a multistoried agricultural plot, mimicking naturally tiered tropical canopies and ground covers. Erosion from tropical downpours is thus minimized. In reproducing the species diversity of the original forest cover, this farming method transforms a rainforest into a food forest.

In the semiarid tropical regions of Africa, the sharp demarcation between wet and dry seasons promoted the domestication of suitable crops and the development of new farming strategies. Pearl millet and sorghum were revolutionary crops, perfectly adapted to the wide seasonal swings of temperature and moisture. Paramount among agricultural practices was the combination of these drought-tolerant cereals with nitrogen-fixing legumes to provide a full complement of protein. Food availability in semiarid Africa was moreover enhanced by flexible land-use practices that encouraged different ethnic specialist groups to cooperate for mutual benefit. In areas that would support agriculture, farm plots were turned over to herders at the end of the wet-season harvest. The livestock would occupy the farmland as pasture; as the animals grazed the cereal stubble, they deposited manure, a process that continuously renewed soil fertility and so prepared the plot for its return to cultivation. This complementary land-use system contributed to the diverse food landscapes that Africans developed over millennia in a tropical continent not especially favored with fertile soils.

With such techniques, African farmers transformed tropical environments into harvestable food plots. The agricultural principles used to accomplish this are not self-evident or easily acquired. They are inherited as cultural funds of knowledge. The legacy of African domestication is most clearly seen in tropical landscapes, with influences that would traverse oceans and continents.

African Plants on the Move

Sankofa is the mythic bird that flies forward while looking backward. The egg
it holds in its beak symbolizes the future. Sankofa's flight reminds us to look
to the past in order to move forward to the future.

Akan folklore

Unumbotte then gave sorghum to Man, also yams and millet. And the people
gathered in eating groups that would always eat from the same bowl, never
the bowls of the other groups. It was from this that differences in language
arose.

Bassari legend

THE CONTINENT OF AFRICA WAS anything but peripheral to the vast trading networks that connected peoples of the ancient world across land and sea.
Africa intersected the maritime routes that initially linked Asia to the Mediterranean. Plants and animals that came out of Africa through these routes are
rarely appreciated as African domesticates. The donkey, for instance, became
the dominant beast of burden throughout much of the Old World. It appears
throughout the Old and New Testaments and is depicted in Egyptian hiero-
glyphs. When Carthaginian Hannibal crossed the Italian Alps during his campaign against Rome, he did so with African forest elephants (a species smaller
than its savanna cousin, now extinct in its former North African range).[1] There
was even a demand for the continent's wild animals, which were frequently cast
in Roman combat spectacles. Africa's participation in intercontinental trade networks grew during the Middle Ages, when Africa supplied ivory (for musical
instruments and board games), gum arabic as a medicinal, melegueta pepper
(or "grains of paradise") for seasoning dishes, and gold for coinage. But food
too was a part of the ancient trade in commodities. Thousands of years ago,
African food crops journeyed beyond the areas of their domestication. Sorghum,
pearl and finger millet, and several legumes were introduced to other parts of

the Old World. There the African crops revolutionized food availability in semi-arid landscapes.

Although the Columbian Exchange begins its historical timeline with the fifteenth-century maritime expansion of Iberians, significant botanical transfers also occurred in the pre-Columbian period. Notable are the Asian crops that diffused with the expansion of Islam in the seventh century, particularly sugarcane. Plants were also transported through ancient Indian and Pacific ocean networks, whose seafarers possibly connected Asia and Oceania with the Pacific coast of the Americas.[2]

Botanical texts strive to determine the geographical areas where crops originated. Scholarship on transoceanic and intercontinental crop exchanges attempts to shift the focus by identifying the plants that accompanied human beings across oceans and lands. However, neither perspective places subsistence, and the trading systems supported by subsistence, at the center of concern. Over the millennia preceding the first visits to Africa of Portuguese mariners in the fifteenth century, African crops journeyed to Asia. Africans additionally made important contributions to the development of several introduced Asian food plants in the same time period. Thus begins the botanical legacy of a continent.

African botanical exchanges with the ancient world are evident in the archaeological and historical record in three distinct eras. They begin three to four thousand years ago through Indian Ocean trading routes to India and beyond. We see them next in the florescence of a desert agricultural kingdom in southern Libya in the first millennium B.C.E., which linked sub-Saharan Africa to the Mediterranean. They are additionally evident with Muslim expansion from North Africa to the Iberian Peninsula from the eighth century.

AFRICAN CROPS AND THE
DESERT KINGDOM OF THE GARAMANTES

Greek historian Herodotus (ca. 484–425 B.C.E.) wrote of Africa as an inhospitable place, the hottest part of the known world, at a time when the Garamantian kingdom in southern Libya was developing new agricultural practices to cope with increasing aridity.[3] The Garamantes were descendants of Berber and Saharan pastoralists, who since ancient times had moved herds of cattle, sheep, and goats across Saharan grasslands. They settled in southern Libya in a region of lakes and springs, where by the beginning of the first millennium B.C.E. they had sedentarized and begun the cultivation of crops. Herodotus viewed the Garamantes as barbarians who dwelled at the periphery of the civilized world; nevertheless, their settlements lay astride the crucial caravan routes that connected the Mediterranean world to sub-Saharan Africa (especially to Gao on the Inland Delta of the Niger River) and the Nile Valley to the east. Between the first millennium B.C.E. and 500 C.E., the Garamantes became a dominant power in the central Sahara, in part because of the trade in salt, slaves, and wild animals to Rome, but also because of an agricultural system that supported settlement in a region that was experiencing rapid desertification. Recent archaeological research has improved our understanding of the methods the Garamantes used to make the desert bloom and of the role of African domesticates in the region's specialized agricultural system. As a consequence, the traditional view of the Garamantes as a desert outpost of Mediterranean food systems is changing to one whose success was also crucially linked to the achievements of farming peoples living in the savannas south of the kingdom.[4]

The Garamantes began the shift to agriculture with a crop repertoire initially inherited from the Mediterranean. They planted emmer wheat, barley, dates, grapes, and figs as foodstaples. But by the first millennium B.C.E., the region's remaining lakes and springs had disappeared. However, the rainwater that once sustained the Garamantes did not entirely vanish. Significant deposits survived underground, with an enormous lake of subterranean water lying beneath the Garamantian heartland. This fossil water had accumulated during a series of wet climatic cycles that came to an end around 3500 B.C.E., when the Sahara shifted to a permanently drier climate. In response to the continuing desertification, the Garamantes began developing a water-extraction system (called foggaras) that took water from the vast underground aquifer.[5]

In the middle of the first millennium B.C.E., the Garamantes made the desert bloom by adopting the foggara technology that the Persians had previously introduced to Egypt. It involved digging vertical shafts to reach the aquifer and channeling the subterranean water flow to a downstream location where it could be captured for irrigation. A prodigious amount of human labor was involved in the construction and maintenance of the irrigation infrastructure. The access shafts were dug some 30 feet apart to reach the water, which was located 30–130 feet below the surface. Eventually 500 miles of underground channels were dug. This water-extraction network supported the food system that made the Garamantes a dominant power over 70,000 square miles of desert in the central Sahara. The cultivated fields irrigated with fossil water enabled the kingdom to achieve a population concentration of perhaps as many as ten thousand people. In the Sahara Desert an urban society had developed on top of the subterranean water that flowed beneath it.[6]

The Garamantes accomplished this feat through slavery, principally with peoples drawn from the dense farming populations south of the Sahara. Initially salt was traded for slaves, but as the Garamantes grew increasingly reliant upon the water-extraction system for their subsistence, they turned to slave raiding to supply the necessary labor force for its maintenance and expansion. Their slaving expeditions, which reached all the way south to the shores of Lake Chad, netted captives from agricultural societies. The presence of sub-Saharan farmers in the Garamantian kingdom coincides with the introduction of two important crops to the foggara food-production system—drought-tolerant sorghum and pearl millet, both originally domesticated in the Sahel. The indigenous African cereals are evident in the archaeobotanical record of the Garamantes during the fourth century B.C.E., at a time when the expansion of the irrigation system was underway. Sorghum and millet contributed considerably to overall food supplies because, in contrast to winter-grown Mediterranean cereals, the African domesticates thrive in the hotter summer months. In enabling a second cycle of agricultural production within the calendar year, the African domesticates made a substantial contribution to the kingdom's cereal reserves. An additional food harvest during the summer would have been critical to subsistence availability at the peak of foggara development between the first and fourth centuries C.E., when the estimated workforce likely numbered one to two thousand slaves. The Garamantes' revolution in food production thus was influenced by the arrival of farming peoples from the Sahel, who were enslaved to build and service the irrigation system, but for whom sorghum and millet were traditional dietary staples.[7]

By the fourth century C.E., when slaves formed perhaps as much as 10 percent of the population, the groundwater table that sustained the Garamantes began an irreversible decline. Excavation to reach the aquifer grew more difficult and labor demanding. The fossil reserves proved finite, and around 500 C.E. the kingdom collapsed. The population dispersed to desert oases and mountainous areas with stream flow and there practiced irrigated farming on a smaller scale. Sorghum remained an important crop to their "Berber" mixed-race descendants, as it fed both herders and their livestock.[8]

ISLAMIC EXPANSION AND THE DIFFUSION OF AFRICAN FOOD CROPS

African crops infiltrated the food systems that accompanied the expansion of Islam from the seventh century. Muslim trading networks moved Asian rice, citrus, and sugarcane from India to the Middle East, the Mediterranean, Egypt, and East Africa. But they also transported sorghum, pearl millet, and other plants from the African continent to geographical areas within the burgeoning caliphate.

One of the earliest historical references to cultivation of African cereals in the Muslim world comes from Mesopotamia, where slave-based sugar plantations developed near the caliphate stronghold of Basra in the south of Iraq. Cultivation of these marshlands for sugarcane and other crops required extensive reclamation of soils, an arduous task. Africans, mostly taken from East Africa (known in ancient times as Zanj), comprised a considerable segment of the tens of thousands enslaved to carry out the work. Slaves were made to remove by hand the salt crust, or natron, from the topsoil in order to expose the fertile soil beneath it. Many African domesticates were cultivated for food, including sorghum, millet, and melons.[9]

In a prelude to the conditions that prevailed on sugar plantations in tropical America, Basra's slaves toiled under wretched conditions. Inadequately nourished and unable to improve their social conditions, the Mesopotamian Zanj led two ninth-century slave uprisings. In the major rebellion of 868–69 C.E., they captured Basra and other towns, built their own capital city, and achieved fourteen years of freedom before their defeat and reenslavement in 883 C.E.[10]

The rise and expansion of Islam facilitated the geographical dispersal of several other crops of sub-Saharan origin to the Middle East and Europe. Coffee (the esteemed *arabica* type), which originated in Ethiopia, was already a feature of daily life in the Arabian Peninsula and Middle East before the Dutch popularized the beverage in Europe during the seventeenth century. The expansion of Islam across North Africa, along trans-Saharan caravan routes and southward to African savanna societies, invigorated ancient long-distance trade networks with the Mediterranean. It contributed to the northward diffusion of the West African kola nut, a stimulant even richer in caffeine than coffee, and the appearance of melegueta pepper and gum arabic in medieval European markets.[11]

African sorghum transformed food production in Muslim Spain. Repeated migrations of Berbers from Morocco to al-Andalus between the eighth and eleventh centuries led to its diffusion to the Muslim-held Iberian Peninsula.[12] Sorghum was ideally adapted to Berber agropastoral practices, as it doubled as a food and feed crop.[13] The drought-tolerant grain produced a second harvest in the Mediterranean climate during the hot and dry summer and became the dietary staple of the lower classes of Islamic Spain. It was prepared in a thick porridge resembling polenta in consistency and use. Sorghum so changed subsistence patterns that Ibn Khaldun, who lived in fourteenth-century Seville, attributed the good health of the population to its "diet of sorghum and olive oil."[14] A Latin copy of an eleventh-century medical text by Ibn Butlān (d. 1038) illustrates a field of sorghum (figure 2.1).

FIGURE 2.1.   Illustration of sorghum, eleventh-century Spain, by Baghdad physician Ibn Butlān (d. 1038), from Latin translation of his treatise *Tacuinum sanitatis,* printed late fourteenth century in Italy.

SOURCE: Reprinted with permission of Österreichische Nationalbibliothek, Ser. Nov. 2644, fol. 48v.

THE MONSOON EXCHANGE

The African millets are evident in South Asia at a much earlier date. They found a particular niche in the agricultural systems of the arid and semiarid tropics, where they provided an adaptive cereal. The archaeological record from western India indicates the presence of sorghum and millet between 2000 and

FIGURE 2.2. Monsoon Exchange, 2000 B.C.E.–500 C.E.

1200 B.C.E. Finger millet, cowpea (also valued as cattle feed), and the lablab or hyacinth bean reached South Asia during the second millennium B.C.E.; the tamarind, pigeon pea, okra, and the castor bean followed.[15]

The seeds of the African food plants made their way past the Arabian Peninsula to the Indian subcontinent along two principal trade routes: one that linked the Ethiopian Highlands to the Horn of Africa; the other connecting the East African highlands to Zanj ("Land of the Blacks," now known as the Swahili Coast) (figure 2.2). The coastal ports formed the African legs of a maritime trading network that spanned the Indian Ocean. Recovered artifacts suggest that East Africa and Asia were trading as early as 3000 B.C.E.[16] The interconnected bodies of water were known as the Erythraean Sea. The navigation route from Roman Egyptian ports and the trading opportunities found along the way are described in the navigational guide, the *Periplus of the Erythraean Sea*, written in the first century C.E. The patterns of the seasonal winds enabled ancient mariners to sail across the Indian Ocean. Between December and March, the monsoon blows from the northeast, which permits downwind voyages westward to the Red Sea. The reversal of wind direction between June and September facilitated maritime crossings to India. By sailing the monsoon winds, mariners could complete the circumnavigation of the Erythraean Sea in just one year.

Recognizing the significance of the wind-assisted trade networks across the Old World tropics for the plant transfers that occurred between Africa and India between the second millennium B.C.E. to first millennium C.E., historian J.R. McNeill terms this period of ancient crop movements the Monsoon Exchange.[17] Notable among the Asian plants brought to Africa by the Monsoon Exchange are two tropical root crops, taro (known also as cocoyam) and the banana. Both are of Southeast Asian origin. They became important foodstaples of the African humid tropics over the same period that sorghum, millet, the *Vigna* species of legumes, and watermelon journeyed eastward to semiarid India.

Botanical studies place domestication of the banana's wild progenitors in Southeast Asia, although a second center is suggested by the presence of edible cultivated bananas in New Guinea possibly seven thousand years ago. Wild bananas contain many seeds and little comestible pulp. Domestication to a more edible fruit increased the ratio of pulp to seed, a process that eventually left the plant entirely seedless. But selection for the seedless trait meant that the banana was no longer able to reproduce by natural means. It could only be propagated by the removal and replanting of vegetative cuttings of the offshoots (or suckers) that develop from the underground stem (or corm) of the mature plant. Domesticated bananas thus depend upon people for reproduction. Human agency was moreover necessary for the plant's dispersal throughout Asia and beyond.[18]

The banana arrived in Africa through the maritime networks of the Monsoon Exchange. It accompanied seafarers from Asia across the Indian Ocean to the African continent. The plant's value to ancient navigators was not as we might suppose, for its fruit, but rather its underground starchy stem, which is also edible and remains so for long periods—the perfect victual for extended voyages at sea. Furthermore, cuttings of the stem stored for several months will not lose the capacity to develop into normal plants if they are later replanted. There is little doubt that *Musa* root stems, with their long-term regenerative capability, were introduced to Africa through westward monsoon voyages.[19]

The term "banana" actually refers to two closely related edible plants of the *Musa* genus, the banana and plantain.[20] While the Western world categorizes the two by calling the sweet "ready to eat" fruit the banana and the starchy cooking species the plantain, this distinction is not observed in many tropical societies, where cultivars of both types are cooked. Africans prepare plantains and bananas in a variety of ways. They boil, steam, poach, bake, and pound the starchy staple as an accompaniment to meat and vegetable stews. Some cultivars are eaten as fruit; others are brewed into beer. The plant's leaves are used for wrapping food; its underground rootstocks are edible and provide animal fodder. The practice of eating the stem is observed in equatorial Africa, where it is sometimes consumed as a famine food.[21]

In Africa, plantains are generally grown in lowland equatorial forests, bananas at higher tropical altitudes.[22] The geographical focus for plantain cultivation centered on the rainforests of West and Central Africa (from southern Cameroon south and east through the tropical rainforest), whereas bananas came to dominate the East African highlands or Great Lakes area of East Africa. Linguistic evidence for *Musa* cultivation in Africa dates to the first millennium C.E., when the plant entered the continent by way of East Africa.

Despite the proximity of East Africa to the ports of the Monsoon Exchange, early evidence for *Musa* in Africa comes from an archaeological site in southern Cameroon. Although there is still debate about whether the plant remains are in fact *Musa,* the phytoliths recovered from the bottom of ancient refuse pits have been dated between 840 and 370 B.C.E.[23] If correct, this would date the Asian plant's cultivation by Africans on the western side of the continent to the last millennium B.C.E.[24]

The remarkable advantages of the *Musa* plant attest to its ready adoption by African farmers as a foodstaple. It is among the highest-yielding crops, and the labor inputs are minimal. In contrast to cereals, bananas and plantains can be harvested throughout the year, thus providing dependent populations a constant and reliable food source. Yields of the plant are extraordinary: over 200,000 pounds can be taken from a single acre. This figure is ten times the yield of yams and one hundred times that of potatoes grown on an equivalent amount of land. An established banana garden will produce for thirty or more years, nearly the entire working life of an average African farmer.[25]

The importance of the banana to Africans is reflected in the profusion of cultivars developed on the continent. Experimentation and innovation with the plantain and cooking banana created a rich diversity of new types. As a consequence, there emerged secondary centers of banana diversification outside Asia—one with plantain in the west-central African rainforest, and the other with the highland cooking banana in the region of Lake Victoria. Today we can recognize some 120 plantain and 60 banana cultivars that were developed in these two areas. This could not have been easily accomplished. Vegetative reproduction ensures the exact transfer of the parent's genetic material to progeny. Africans could only have realized the remarkable array of varieties through patient observation and selection of desirable traits for cloning. Genetic studies have estimated the mutation frequencies at which the African cultivars accumulated their distinguishing characteristics. The research suggests it would have taken some two thousand years for the African plantains and perhaps half that for the highland bananas.[26]

In providing tropical Africa a productive dietary staple, the banana plant encouraged an agricultural transformation of the African humid tropics. Banana cultivation made significant contributions to population growth and, by releasing more people from farming, facilitated new kinds of trade relations and political centralization in and around the Congo Basin and the Great Lakes region of East Africa.[27] The history of the banana plant in Africa illuminates the agricultural skills and innovative practices that African farmers mastered and continuously developed in the millennia prior to the arrival of Europeans.

THE PLANTAIN'S EXPANSION INTO THE ATLANTIC

The plantain and banana were thus well-established African dietary staples when Portuguese navigators ventured down the continent's Atlantic coast in the first half of the fifteenth century. Portugal's motive was to find direct gateways to the fabled sub-Saharan source of the gold trade that crisscrossed North Africa.[28] However, different commercial interests beginning to form off northwest Africa would soon give new impetus to Portuguese activities on the adjacent mainland. The movement of the plantain/banana from Africa into the Atlantic world is part of that story.

The early migration of plantains and bananas begins with the fifteenth-century maritime expansion of Portugal and Spain to two island archipelagos off the coast of Morocco, the Madeira and Canary islands. Settlement was driven by a new kind of "white gold," sugar, demand for which would propel the burgeoning Atlantic economy. Sugar was a revolutionizing commodity. It demanded considerable labor to tend and cut the fields and feed the cane through the sugar mills, tasks that over time increasingly fell to enslaved Africans. But this workforce also required nourishment—basic foodstaples necessary for human survival. The plantain's arrival on the Atlantic sugar islands coincided with the importation of enslaved Africans for whom it was a longstanding dietary preference. The plantain/banana first comes into historical view on the Canary Islands, whence clones were introduced to Spanish Santo Domingo. But the plantain was assuredly a frequent passenger on the Portuguese caravels that transported African slaves to these island chains.

The Madeira Islands were the first prominent site of European sugar outside the Mediterranean.[29] For the Portuguese who discovered the island archipelago in 1418, it had the virtue of being the closest of the Atlantic islands to the Iberian Peninsula. The Madeiras lie approximately 350 miles west of Morocco. Settlement of the principal Madeira island was underway by the 1420s, and despite the profitable felling of its vast forests for timber exports ("Madeira" is Portuguese for "wood"), a superseding interest in agriculture resulted in complete deforestation by the end of the decade. Introduction of sugarcane quickly followed, with exports to England recorded in 1456.[30] Madeira soon became the single largest sugar producer in the Western world, anticipating by half a century the plant's diffusion to the New World.[31]

Cane cultivation on Madeira Island itself was supported by a connected system of irrigation aqueducts, conduits, and tunnels designed to deliver mountain water to agricultural fields below. Eventually spanning a distance of over four hundred miles, construction of the *levadas* was a remarkable though dangerous and back-breaking feat on an island only thirty-seven miles long. Nonetheless, it served the nascent sugar industry well by providing a reliable water delivery system that supported its rapid expansion.[32] In this early period, sugar cultivation used different forms of labor, some of it paid, others enslaved. Slaves included Guanche captives native to the Canary Islands, Moors from Morocco, and increasingly after the 1440s, sub-Saharan Africans.[33] By the 1480s, the Madeiras' sugar expertise had transferred to the Canary Islands, the archipelago located to the south. The expansion of sugar cultivation inevitably hastened the demand for slave labor.[34]

The island chain of the Canaries is situated off the coast of southern Morocco. In contrast to the Madeiras, the Canary Islands were inhabited by a people known as the Guanches when fourteenth-century European cartographers marked the islands on their maps.[35] Portugal used the islands as a way station for collecting wood and water for voyages along the African coast. Decimation of the native Guanche people began in 1402 with the conquest of Lanzarote, the island nearest the African mainland. Those Guanches not killed were enslaved. A protracted military conquest of each island continued over much of the fifteenth century, despite fierce Guanche resistance and continuing disputes between Spain and Portugal over the archipelago's ownership. In 1479 the Treaty of Alcaçovas ceded the Canaries to Spain. In return, Spain acknowledged Portugal's claims to the Madeiras, Azores, Cape Verdes, and trading posts on the African littoral.[36] The Spanish defeated the last of the Guanche insurgency on Tenerife in 1496. Decimation and enslavement of the remaining native peoples cleared the way for sugarcane. On Grand Canary Island, it was planted immediately following pacification of the Guanches in 1483.[37] Spain thus followed Portugal's lead in establishing sugar in the Atlantic islands.

Within ten years, Christopher Columbus had begun his series of historic voyages to the Americas, and he stopped in the Canaries during each outbound journey. By then a sugar economy based on slave labor was in full flower. Columbus, in fact, introduced sugar to Santo Domingo with cane taken from the Canary Islands during his second voyage in 1493. He was already quite familiar with sugar; Columbus had apprenticed in the sugar trade on Madeira fifteen years earlier in 1478.[38]

As the sugar economy of the Madeiras and Canaries grew in the second half of the fifteenth century, so did the demand for slaves to work the cane fields and mills. The supply of Guanches, some of whom were exported to Madeira during the founding decades of Atlantic sugar production, diminished as Spain exerted full control over the Canaries (the Guanches were reported to be nearly extinct by 1540).[39] The demand for slaves led the Portuguese and Spaniards to the African mainland. The easternmost Canaries—Lanzarote and Fuerteventura—are, for instance, located just sixty miles from mainland Morocco. Spanish raids along the Moroccan coast supplied the Canary Islands with Berber slaves, livestock, and camels.[40] Meanwhile, Portuguese expansion to the African mainland had begun as early as 1415 with the taking of Ceuta across the Straits of Gibraltar. Ceuta supplied slaves to Lisbon and the Madeiras, as did slave raids along the African coast.

The geographical focus of the slave trade shifted decisively in the 1440s, when Portugal established its first trading post at Arguim, an island adjacent to Mauritania and within reach of coastal peoples, especially the dense populations found south of the Senegal River (Rio de Canagua).[41] Figure 2.3, an early sixteenth-century map of this stretch of the African coast, shows the slave port's proximity to the livestock-herding Azenegues, who lived north of the Senegal River and whose proximity to Arguim made them vulnerable to slave raiders.[42] The trade in enslaved Africans gained momentum as Portuguese navigators extended the metropole's reach with journeys south along the Guinea coast. By 1460 they were in Sierra Leone; the equator was crossed in 1475, and the Congo

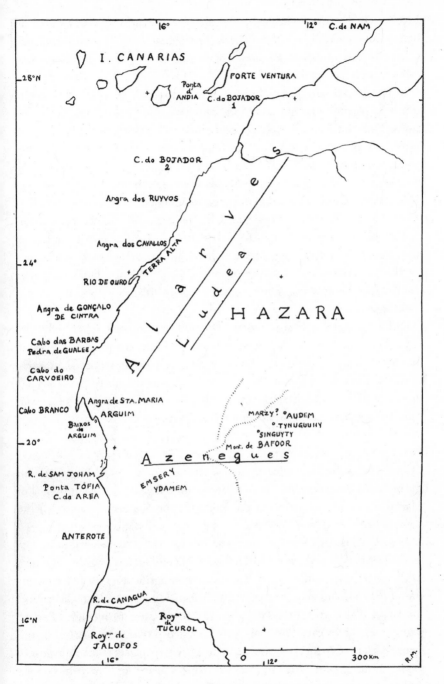

FIGURE 2.3. Map of region north of Senegal River, early sixteenth century.

SOURCE: Pacheco Pereira, *Esmeraldo de situ orbis,* opposite p. 24.

River reached by 1483. The Portuguese established a sugar colony on the island of São Tomé in 1485. From the 1470s, Portuguese mariners were in repeated contact with Bantu-speaking peoples of west-central Africa for whom the plantain was a dietary staple.[43]

In a rehearsal for the greater historical drama to come, enslaved Africans were steadily replacing Guanches and Moroccans in the sugar fields of the Atlantic islands. The ethnic shift in the sugarcane workforce over the second half of the fifteenth century coincided with the appearance of the plantain in the archipelagos. The historical record credits the Portuguese with introducing the banana plant to the Canaries sometime between 1402 and the 1480s, when ownership of the island chain was in dispute and conquest of the Guanches underway.[44] More likely the plantain arrived after midcentury, when Portuguese caravels began carrying slaves from African societies long familiar with its cultivation and consumption. As with sugar, the history of the banana plant in Spain's New World colonies begins at the gateway of the Canary Islands. Dominican friar Tomás de Berlanga took clones from the Canary Islands with him to Santo Domingo in 1516.[45]

Although Spain established a toehold in southern Morocco (a disputed polity today known as Western Sahara), the Treaty of Tordesillas in 1494 effectively divided the known world between Portugal and Spain. In return for granting much of the new territory of the Americas to Spain, the treaty effectively left most of Africa under the control of Portugal. One important consequence was that Portugal became the initial supplier of enslaved Africans to Spanish America until its monopoly in the transatlantic slave trade was contested by other European powers in the seventeenth century.[46]

Spaniards and other Europeans often wrote of "figs" in initial encounters with the unfamiliar banana plant. The Italian Antonio Pigafetta, chief chronicler of Magellan's circumnavigation of the globe for the Spanish crown, filled the nomenclature void by using the word "fig" to describe the crew's first experience with the banana in the western Pacific in 1521.[47] Huguenot Jean Barbot used the same name for the African foodstaple on his slave voyage to Guinea in the late seventeenth century. He drew it alongside another plant he encountered, the maniguetta pepper tree or melegueta pepper—the medicinal and spice that had been carried by trans-Saharan caravans to medieval Europe (figure 2.4).[48] However, Portugal's early presence in western Africa during the formative period of the Atlantic slave trade likely explains Portuguese exposure to and adoption of the African common name "banana" used in many Guinea coast languages. Garcia da Orta, a Portuguese physician and botanist who traveled from Portugal to India in 1534, made an explicit reference to the "figs in

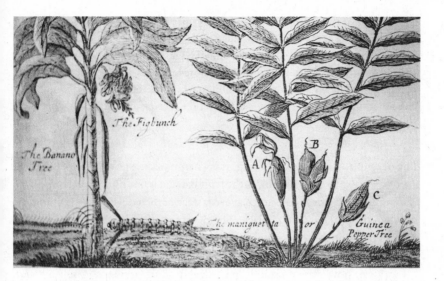

FIGURE 2.4.   Illustration of banana and maniguetta (melegueta) pepper tree, late seventeenth century.

SOURCE: Barbot, "Description of the Coasts of North and South Guinea," in Churchill, *Collection of Voyages and Travels,* vol. 5, pl. F, opposite p. 128.

Guinea, which they [Africans] call bananas."[49] In his journeys to Congo and Angola in 1578–79, the Portuguese Duarte Lopez used the African language name to refer to this important foodstaple: "a great quantity of fruit is found here, named bananas by the natives."[50]

In the early sixteenth century, Valentim Fernandes described the banana as "the best thing to eat" on the African island of São Tomé. Dutch mariner Pieter de Marees wrote of its pleasant taste on his visit to Lower Guinea in 1600. However the first "banana" introduced to the Madeira Islands was more likely the African plantain, which became known there as *banana da terra* in Portuguese. The same term was used to reference the cooking banana in sixteenth-century Brazil.[51] Thomas Nichols, who enumerated several introduced crops he saw on his visit to Madeira and the Canary Islands in 1526, used the Spanish word *plátano* to describe his first encounter with the plantain:

> This Island hath singular good wine . . . and sundry sorts of good fruits,
> as Batatas, Mellons, Peares, Apples, Orenges, Limons, Pomgranats, Figs,
> Peaches of divers sorts, and many other fruits: but especially the Plantano
> which groweth neere brooke sides, it is a tree that hath no timber in it, but
> groweth directly upward with the body, having marvelous thicke leaves, and

every leafe at the toppe of two yards long and almost halfe a yard broad. The tree never yeeldeth fruit but once, and then is cut downe; in whose place springeth another, and so continueth. The fruit groweth on a branch, and every tree yeeldeth two or three of those branches, which beare some more and some lesse, as some forty and some thirty, the fruit is like a Cucumber, and when it is ripe it is blacke, and in eating more delicate then any conserve.[52]

The description of the plant being ripe when its skin was black indicates that the reference is to the plantain, as the fruit banana is not usually eaten this way. Plantains, which can be consumed at every stage of their development, are sweetest when the peel has turned black. The sweetness of the ripened plantain contributes to our inability to pinpoint when true fruit bananas were first brought from Africa to the Atlantic islands and the Americas.[53]

European powers that established a slave-trading presence in sub-Saharan Africa eventually all adopted the word "banana" (or "banano") for a food crop that was increasingly important to the commerce in human beings. The Spanish, effectively denied footholds in tropical Africa by the Treaty of Tordesillas, referred to both the cooking and fruit banana in the Canary Islands and Spain as *plátano*. Peter Martyr, an Italian who served in the Spanish court prior to his death in 1526, puzzled over use of this word for a plant that bore little resemblance to the "plane tree which is in no way related to it." But Martyr did aver that the *plátanos* of Hispaniola originated in "that part of Ethiopia which is commonly called Guinea."[54]

Portuguese adoption of the African word "banana" parallels adoption of the plant as a foodstaple on slave ships and in the Atlantic islands. As the sugar economy there increasingly turned to enslaved African labor, a longstanding African dietary staple simultaneously gained a subsistence role in the Atlantic archipelagos. This occurred long before Portuguese navigators reached the Asian regions where Magellan's crew had encountered the "fig" in 1521. The plantain emerges from the background of the Atlantic's first sugar islands as a principal foodstaple. The diffusion of the plantain from Guinea to the Madeira and Canary islands signals an early instance of the growing significance of food to the emerging Atlantic economy and of the reassertion of African dietary preferences among the continent's deracinated slaves.

The plantain was not the only African food crop to become a passenger on Portuguese vessels during the late fifteenth century. Rice also makes an early appearance, both in the first reports of Portuguese explorers visiting Atlantic

Africa, and later in the cargo manifests of vessels returning from there to Lisbon. Alvise da Cadamosto explored the Senegal and Gambia rivers in 1455 and 1456. Observing the Gambia's Mandinka, he wrote "they have more varieties of rice than grow in the country of Senega."[55] The rice Cadamosto encountered was the African species *Oryza glaberrima,* and he provides one of the earliest accounts of European interest in this indigenous food crop. A vessel carrying the grain arrived in Lisbon from the Guinea coast in 1498—a year before Vasco da Gama completed his epochal return voyage from rice-growing India. The earlier date supports the likelihood that this rice delivery to Portugal was the African species.[56]

Plantains and rice are two crops that are typically claimed as examples of the Columbian Exchange, as plants Portuguese mariners introduced from Asia. However, each was a significant and longstanding dietary staple in Atlantic Africa when the Portuguese first ventured southward along the continent. In the decades before caravels reached the Indian Ocean, the plantain had been adopted as a subsistence staple in Madeira and African rice imported to Lisbon. Where there were African foodstaples, there were enslaved Africans, and in the fifteenth century enslaved Africans were present in significant numbers in the workforce of Lisbon and the Madeiras.[57] Muslim expansion into Africa in the seventh century similarly encountered societies for whom plantains, rice, and sorghum had been dietary staples for thousands of years. African accomplishments in plant domestication and crop development are not made explicit in either expansionary movement to the continent.

Iberian incursions along western Africa brought them in contact with a diversity of cereals and tropical tubers, some entirely novel and others previously known through Muslim introductions. The principal African foodstaples that the Portuguese encountered at the beginning of the transatlantic slave trade are depicted in figure 2.5.

Columbian Exchange scholarship that attributes intercontinental crop transfers to Iberians assumes that Iberians also had a deliberate hand in planting and propagating these crops elsewhere. Early written accounts that present African food plants such as the plantain and banana as novelties and new discoveries profess both curiosity and ignorance, revealing little if any prior familiarity with the ways they were grown. Attributions of Portuguese and Spanish agency in establishing these foodstaples in the Atlantic islands and in tropical America divorce the plants from the purpose they served. Ignored is the role of the plantain and other African foodstaples in the provisioning of slave ships, and the motivation that transplanted slaves themselves might have

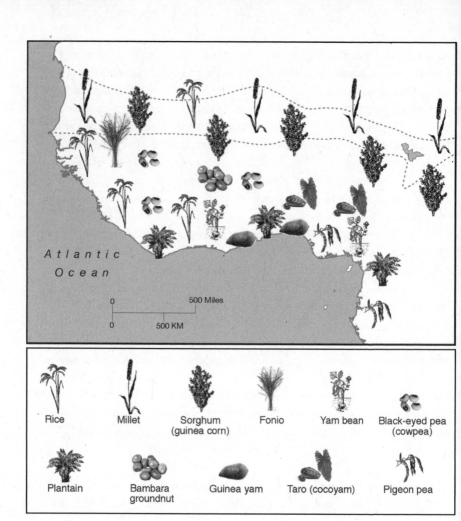

FIGURE 2.5. Principal African foodstaples on the eve of the transatlantic slave trade, ca. mid-fifteenth century.

had in initiating cultivation of longstanding food preferences. By neglecting fundamental consideration of a people historically active in the domestication and breeding of plants and animals, and active even under conditions of bondage, the classic literature of the Columbian Exchange unwittingly contributes to the perception that Africa and its peoples were inconsequential in botanical history.

Alfred Crosby in his book *Ecological Imperialism* drew attention to the role of anonymous immigrants in the intercontinental transfer of species and the shaping of newly encountered landscapes into so-called Neo-Europes. But we should also consider another chapter in the history of intercontinental transfers. In this instance, the migration was not voluntary, but forced; the immigrants were not European, but African and enslaved; and the plants involved were tropical species from Africa. These African plant transfers were unlike any other discussed in the Columbian Exchange literature, for they occurred as vital supports of the transatlantic slave trade.

THREE

African Food Crops and the Guinea Trade

The importance of American foods in Africa is more obvious than in any other continent of the Old World, for in no other continent, except the Americas themselves, is so great a proportion of the population so dependent on American foods. Very few of man's cultivated plants originated in Africa. . . . and so Africa has had to import its chief food plants from Asia and America.

ALFRED W. CROSBY
Columbian Exchange

The ability of Atlantic merchants to export slaves depended directly on their ability to feed them from the moment of purchase until they departed for the Americas.

JAMES E. SEARING
West African Slavery

THE INSTITUTIONAL APPARATUS THAT CUMULATIVELY removed millions of Africans from the continent depended critically on the availability of food. First, the African militias and raiders who took slaves had to be fed. Likewise, so did the Africans who awaited deportation to the New World. The resident slave traders and European officials stationed at forts along the Guinea coast also depended upon food availability. Finally, the slave ships that carried Africans on the Atlantic crossing required provisions enough for journeys that lasted two or more months. The feeding of these multitudes—all involved in the globalizing economy by consent or coercion—depended vitally upon food, even if what served as daily sustenance varied starkly between each set of participants. The subsistence of each of these groups takes us to the underbelly of the Atlantic slave trade and the ways Europeans used the plants domesticated by tropical peoples to support it.

But subsistence signifies more than the food that fed commerce. It asks us to engage the broader relationship of food to culture, and culture to identity. Subsistence connotes the ways that enslaved Africans used the production of food to resist imposed diets and to exercise choices over what served as their daily sustenance—in essence, to win back a modicum of control over their own bodies. Subsistence illuminates the symbolic meaning of specific foods and foodways in different geographical and social contexts and the power relations that inform it.

When slave ships arrived along the African coast, they carried only partial stores of foodstuffs. Most of the food needed to provision human cargoes was grown in Africa. The ability of African middlemen and European ship captains to export slaves depended upon providing them subsistence over the months that spanned their capture and subsequent delivery to the auction blocks of the Americas. African-produced food was pivotal to the harvest of slaves for the Atlantic economy, and the entire commerce in human beings came to rely upon subsistence staples that grew in the tropics of two continents.

Europeans also drew upon food grown in Africa to provision their military garrisons and commercial agents. Part of the reason for this transition away from traditional European foodstaples was the rapid realization that temperate-zone crops were poorly suited to the growing conditions of tropical Africa. As early as 1455, the explorer Alvise da Cadamosto observed during a visit to the Senegal River that attempts to grow wheat, barley, rye, and grape vines had met with complete failure.[1] When the Kongo court converted to Catholicism in 1485, the king asked Portuguese officials to send along with priests, "farmers to teach them to till the soil with plows, and women to teach the baking of bread."[2] But efforts to prepare the sacramental wafer from wheat, a crop indispensable to Catholic liturgical practice, had to be abandoned. The Mediterranean cereal did not thrive in equatorial Africa, nor did draft-animal traction prove a suitable technique for cultivating rainforest soils.[3] African pearl millet and rice provided wheat-flour substitutes for priests, traders, and administrators based along the Guinea coast, while manioc flour eventually served the same purpose for the Portuguese resident in equatorial Africa.

As the Portuguese made their way southward along the African littoral, they encountered diverse agricultural and agropastoral food systems. Along the coast of Upper Guinea, they found rice, millet, sorghum, the cowpea, and abundant livestock herds. Toward the equator, root crops (yams, plantains, and taro), pigeon pea, and the Bambara groundnut (*Vigna subterranea*) dominated indigenous food systems. Most of these African foodstaples were entirely new to the Portuguese. The subsistence staples were all loaded on slave ships, along with transplanted Amerindian domesticates, as provisions for the enslaved Africans the vessels carried.

An examination of the role of Africa-grown foodstaples in the transatlantic slave trade and their diffusion to New World plantation societies illuminates some limitations of Columbian Exchange scholarship, which fails to place subsistence within the context of the Atlantic economy's burgeoning demand for enslaved Africans. The credit for establishing in other parts of the world food that Africans grew when slavers arrived at the continent's shores is given to Europeans, who for the most part did not consume the African crops as subsistence staples, much less know how to grow them.

SUBSISTENCE AND THE EXPANDING EUROPEAN PRESENCE ALONG THE AFRICAN ATLANTIC

When Portuguese mariners began their journeys along the West African coast in the fifteenth century, they gave names to the regions that held their interest. These designations stuck and persisted beyond the entry of other European nations in the Africa trade. Some of these geographic indicators later became the formal names of European colonies, as was the case with the Gold Coast, the Ivory Coast, and Cameroon. The place names identified critical resource zones where Europeans found desirable trade items. The Grain Coast, for instance, was initially named after the "grains of paradise," or melegueta pepper, which Europeans used as a spice and medicinal. A trade in elephant ivory and gold inspired the toponyms Ivory and Gold coasts. The long stretch of the Atlantic coast, from Ghana eastward to the Niger Delta in Nigeria, was called the Slave Coast, thus making more explicit the purpose of the growing European presence. One of the major ports of Nigeria's oil-rich delta, which once sent thousands into slavery and now exports petroleum, is still known as Escravos, after the Portuguese word for slaves.

African locations known for the availability of surplus food were also given descriptive place names. One of these was the Rice Coast, the stretch of the Atlantic coast between Sierra Leone and Liberia, where the grain was routinely available for purchase.[4] The Banana (or Plantain) Islands, located off the coast of Sierra Leone, took their name from the availability of this important African foodstaple. Cameroon was similarly named after the abundance of shrimp (*camarão* in Portuguese) found in coastal estuaries.[5] These place names draw attention to the significance of known food supplies for European navigation and commerce along Africa's Atlantic coast.

Demand for food increased with the growing Portuguese presence in Africa. Soldiers, priests, and administrators needed reliable sources, as did captains of slave ships, who required provisions for their captives. In 1519 the Portuguese

Crown encouraged the establishment of slave-based plantations to ensure food availability along the African littoral.[6] The Portuguese garrison, São Jorge da Mina (later known as Elmina), built on the Gold Coast in 1482, depended in part on locally produced food. A soldier's daily food ration in the sixteenth century included four loaves of dark bread, typically made with flour from African pearl millet instead of imported wheat, which quickly moldered in the tropics. Resident traders, priests, and military personnel otherwise depended on supply ships from Lisbon for foods culturally in demand or needed for Catholic liturgical practices (wheat flour, olive oil, wine), but local substitutions were often made. Africans supplied the fresh fruits and vegetables, live animals, and supplemental foodstaples demanded by Europeans.[7]

On his visit to the Gold Coast at the beginning of the seventeenth century, Dutch merchant Pieter de Marees described and illustrated the thriving trade in food at the African market that developed next to the Mina garrison. Female vendors were especially important to the marketing of foodstuffs, which included vegetables, fruit, and prepared food (such as the fermented porridge called *kenkey*). Figure 3.1 depicts the women arriving at the market, carrying the trade goods on their heads. One can see the variety of African foods that were for sale: millet, sorghum, African rice, plantains/bananas, vegetables, fruits, cooked food, poultry and eggs, in addition to firewood and drinking water.[8]

The Portuguese presence along the African coast established the pattern of dependence upon African societies for surplus food that other European nations emulated when they built forts and trading posts to facilitate the Atlantic slave trade. This was evident from Senegambia south to Angola. By the end of the seventeenth century the Portuguese, Dutch, English, Danes, Swedes, and Prussians had constructed nearly fifty forts (castles) and trading posts along a mere three hundred miles of coastline of the Gold Coast (figure 3.2). African villages, towns, and markets developed in close proximity. The escalating trade in slaves encouraged the concentration of new populations of traders and slaves along the seaboard and a growing demand for surplus food.[9]

Historian Ray Kea documents the shift in Gold Coast settlement patterns during the seventeenth century and the concomitant food demand that accompanied the Atlantic slave trade. The Dutch trading post at Mori grew from a village of 200 in 1598 to a population of 1,500 in 1612; by the end of the eighteenth century, the town numbered 5,000–6,000 inhabitants. Similarly, the Axim settlement near the Ankobra River in the western Gold Coast held a population of 500 in 1631; some sixty years later, it included between 2,000 to 3,000 persons. The Portuguese Mina garrison, which became Dutch-controlled Elmina in 1637, hosted a population of 15,000 to 20,000 in the late seventeenth

Legend:

A. Residence of market governor
B. Granary for storing cereals
C. Market for bananas and fruits
D. Palm wine stall
E. Poultry market
F. Fish market
G. Women selling firewood
H. Women selling rice and *millie* (sorghum and millet)
I. Women selling freshwater
K. Sugarcane

L. Foreign cloth
M. Women selling prepared food *(kanquies or kenkey)*
N. Fetish table
O. Dutchmen visiting market
P. Guard
Q. Road to seashore trade
R. Women entering market via Mina road with products for sale on their head
S. Road to inland settlements

FIGURE 3.1.   Illustration of Cabo Corso market, Gold Kingdom of Guinea, by Pieter de Marees, ca. 1602.

SOURCE: Marees, *Description and Historical Account of the Gold Kingdom of Guinea (1602),* pl. 4, p. 62.

century, making it the largest European outpost in all of Africa. Kea estimates that coastal towns of the seventeenth-century Gold Coast obtained between 40 and 70 percent of their foodstaples from the market. When Marees had visited the region nearly a century earlier, only the Mina garrison supported a daily African produce market; by the eighteenth century such markets were found near all the European forts and associated African settlements (figure 3.3).[10]

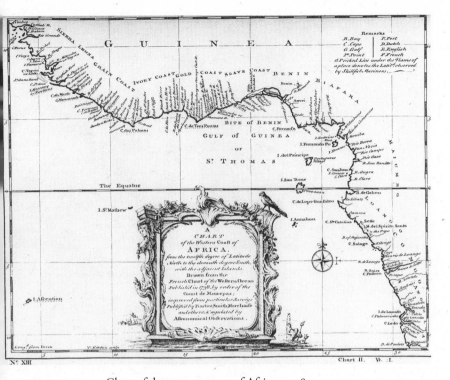

FIGURE 3.2. Chart of the western coast of Africa, 1738.

SOURCE: Reprinted with permission of the antiquarian map collection of the Department of Special Collections, Stanford University Libraries, Stanford, California.

The annual trade in foodstuffs involved thousands of tons of African-grown provisions. Slave-ship captains purchased food along the Guinea coast in various ways: from stocks held by forts operating under the same national flag; from resident European or Euro-African traders affiliated with these outposts; and directly from African and Euro-African merchants in areas of reliable supplies.[11] One Catholic missionary estimated that half the Africans living among the Portuguese and Luso-Africans in Angola were slaves. Many worked on agricultural estates that specialized in the production of subsistence crops. The Jesuit order was likely the largest holder of slaves in seventeenth-century Portuguese Angola, "which by 1658 held 10,000 slaves on fifty plantations."[12] In Luanda, these food farms extended thirty miles inland and supplied both town residents and slave-ship captains with provisions.

The amount of food needed to support a contingent of slaves and their European captors could be considerable. One agent for the French Compagnie des Indes Occidentales at Bissau (Guinea-Bissau) reported in just a nine-day

FIGURE 3.3.   Illustration of the Danish Fort Christiansborg, Gold Coast, attributed to Ludevig Ferdinand Rømer, ca. 1760.

SOURCE: Reprinted with permission of the Royal Library, Danish National Library, Copenhagen.

period that his trading company's activities "consumed 8,200 pounds of rice, 4,200 pounds of millet, 5 cattle, 77 goats, 204 chickens, 400 pounds of flour, [and] 600 pints of wine" from the garrison's supplies.[13] The English Sierra Leone Company dispersed its schooners throughout the coastal region to procure cattle, rice, and other staples directly from African traders.[14]

Seizures of foreign vessels by Dutch ships also illuminate the volume of foodstuffs purchased in Africa for European commerce. In 1719 one captured ship held 12,000 pounds of rice and melegueta pepper in its stores. A cargo of another European vessel confiscated in Lower Guinea carried 34,500 pounds of rice in addition to 8,678 pounds of elephant tusks, 56,700 pounds of melegueta pepper grains, and 320 guilders of gold.[15]

African and Euro-African merchants, who operated as intermediaries between Europeans and African ruling elites, organized the wholesale food trade. They used both overland and coastal trading networks to ensure supplemental demand. Some produced grain and tubers on large estates run with slave labor. The coastal trade was carried out by fleets of canoe barges that transported sizeable quantities of food to the European enclaves and to slave ships anchored offshore. African foodstaples figured prominently in the provisions carried to slaving areas along the Atlantic coast. These included millet, rice, sorghum, plantains, yams, beans, palm oil, and live animals.[16] The live-animal trade involved small ani-

FIGURE 3.4. *About their Barges with which they sail on the sea and how they make them out of a tree*, Gold Kingdom of Guinea, by Pieter de Marees, ca. 1602.

SOURCE: Marees, *Description and Historical Account of the Gold Kingdom of Guinea (1602)*, pl. 8, p. 116.

mals (guinea fowl and poultry) and livestock (sheep, goats, and cattle). Cattle were already an important component of the coastal food trade at the beginning of the seventeenth century when Dutch mariner Pieter de Marees drew them being transported live in canoes (figure 3.4). His drawing of the indigenous West African cattle is perhaps the earliest depiction of the animal by a European.

However, the diverse foodstuffs that Europeans and townspeople purchased for their subsistence represented only a fraction of the total food in demand along the African Atlantic seaboard. A profitable commerce in human beings required sources of surplus food to feed slaves en route to the Americas. From the moment of capture until their final disembarkation, slaves had to be fed. Keeping them alive in transit presumed surplus food availability and sufficient stores to sustain them. Slave ships magnified the demand for food produced in Africa through their purchases of the stores necessary to provision a boatload of captives over the Atlantic crossing—a journey that could take months. It was in this context that the Amerindian domesticates, maize and manioc, were introduced and quickly became prominent subsistence crops in Atlantic Africa.

In the late sixteenth century, west-central Africa—the region south of the

Congo River including Angola—emerged as a major geographical focus for Portuguese slaving.[17] This region represented an ecological transition from an equatorial climate to one of increasing aridity from Luanda southward. In the wetter regions, plantains and the African groundnut dominated food supplies; in semiarid areas, sorghum and millet were cultivated along with indigenous beans. By the 1590s the removal of captives from the African interior to the coast was having a dual effect on food availability. It depleted rural areas of a vital segment of the labor force—mostly youths—thereby disrupting agricultural production. The forced exodus also aggravated demand for food in the ports where slaves were held before embarkation. This resulted in chronic threats of food shortages. Slave merchants could therefore not be assured of finding sufficient African foodstuffs for their ships' stores. By the end of the sixteenth century, rather than taking the risk, Portuguese merchants operating from Brazil stocked locally produced manioc flour (farinha) aboard ships before sailing to Angola for their cargoes of slaves.

Manioc proved an ideal foodstuff because, in contrast to European cereals, its flour does not go rancid and keeps well in tropical conditions. Manioc is a high-yielding source of food calories; it is drought tolerant and resistant to pests. Its principal virtue during the transatlantic slave trade was that, unlike the perishable tuber, its flour could be stored for long periods of time. This made it ideal for protracted sea voyages, but land-based soldiers also depended on it. Portuguese soldiers garrisoned in Angola's slave ports partly relied on imported manioc flour for their basic sustenance. The annual import trade in manioc flour in this early period of the Angolan slave route was quite extraordinary. In the first decade of the seventeenth century, an estimated 680 tons of manioc flour was exported each year from Brazil. The flour was produced by Amerindians, who were made to grate and process it from vast numbers of manioc tubers. Jesuit missions in Bahia also drafted Amerindian labor to prepare manioc flour, which they exported to their religious confreres in Angola in exchange for African slaves.

Agricultural production reoriented over the seventeenth century away from importations of manioc flour to cultivation of the tuber in Angola. Amerindian maize and sweet potatoes improved the ability of coastal ports in Angola to victual slave ships, but traditional staples such as plantains, sorghum, and beans remained important as provisions. Manioc cultivation became self-sustaining once the process of expressing the toxic hydrocyanide compounds from the roots was mastered. The grueling work of processing the tuber into flour for slave ships was assigned to African slaves of little value in the transatlantic trade—chiefly, children and the elderly.[18]

Before the adoption of manioc, maize had already become indispensable to the African agricultural systems that were widely being reorganized along the Atlantic coast to meet slave traders' growing demand for food. Maize provided a high-yielding, quick-maturing cereal that was easily prepared and whose grains could be stored for long periods. The particular success of maize in Lower Guinea, for example, was not due to a dearth of indigenous foodstaples or the incapacity of the region's agricultural systems. This Columbian Exchange crop flourished because its cultivation and preparation made the cereal ideally suited to the trade in human beings.

Maize revolutionized agricultural production in Africa even as it targeted the segment of the population who had been made slaves. It was planted to expedite and ensure delivery of slaves to New World plantation societies. In this sense, maize can be seen as a symbol of the dehumanizing condition of chattel slaves, who were no longer able to exercise dietary preferences or choose the type and amount of food they consumed. A serving of maize was a daily reminder that one no longer held the fundamental right to eat the food that traditionally defined membership in a culture. It signaled the rupture of fundamental entitlements to traditional preferences. Maize emphasized a person's demotion from human being to commodity, the loss of social status and cultural identity, of being made a kinless and orphaned servant in the Atlantic world.

SUBSISTENCE AND THE INTRODUCTION OF MAIZE ALONG THE LOWER GUINEA COAST

Maize was introduced so early to western Africa that some scholars initially believed it was present before Europeans arrived in the fifteenth century.[19] By the 1540s it was planted in the Cape Verde Islands. Plant remains in the archaeological record show that it was grown in the interior of the Gold Coast by the seventeenth century. Maize is the sole Amerindian representative of the useful food plants that Marees depicted on his visit to the Gold Coast at the beginning of that century (figure 3.5).[20] His drawing places maize alongside prominent indigenous African foodstaples—rice, millet, cowpea (black-eyed pea), fonio (*Digitaria exilis*), *néré* (*Parkia biglobosa*), melegueta pepper—in addition to ginger and sugarcane, which had been introduced to the continent before 1500.

Millet, sorghum, yams, plantains, and palm nuts are traditional foodstaples of Lower Guinea. The insertion of maize into the assemblage of African domesticates cultivated in the region had ecological and dietary significance, as archaeologist Ann Brower Stahl points out. In the Gold Coast, maize is harvested during the "hungry season," the lean months of chronic shortfall in be-

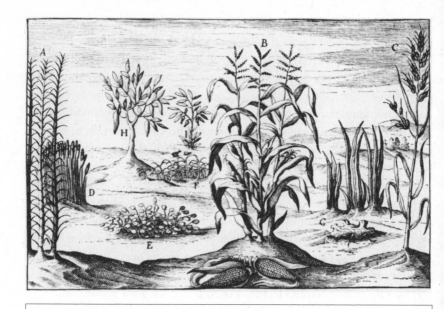

Legend:

A. Sugarcane
B. Maize
C. Rice
D. Millet
E. Cowpea/black-eyed pea
F. Unidentified leafy plant, the size of parsley
G. Ginger
H. Possibly baobab *(Adansonia digitata)*
I. Manigette (melegueta) pepper tree.

FIGURE 3.5. *What Spices and Grains grow in this Country, and what qualities or virtues they have,* Gold Kingdom of Guinea, by Pieter de Marees, ca. 1602.
SOURCE: Marees, *Description and Historical Account of the Gold Kingdom of Guinea (1602),* pl. 13, p. 158.

tween the yam and sorghum harvests. The adoption of maize thus closed a gap in the cycle of agricultural production by facilitating food availability year-round.[21] As the transatlantic slave trade intensified, the need to create surpluses in order to feed its various participants changed the broader consumption pattern of traditional African foodstaples and placed the Amerindian introduction in a prominent role. Seen from this vantage, maize, like manioc, was not, as some accounts would have it, the benevolent and timely rescuer of a chronically hungry continent. These views situate the arrival of Amerindian staples within a framework that considers Africa as a region of few indigenous food crops, whose agricultural systems have long been unable to sustain its peoples. Instead, maize and manioc were expeditious contributors to the Atlantic slave trade's subsistence reserves. Discussions of Columbian Exchange crop introductions to Africa thus should not be divorced from their role in enabling the commerce in human beings. They cannot ignore the transatlantic trade's considerable impact on demand for food grown in Africa.

Over the seventeenth and eighteenth centuries, African millet retained its importance in Lower Guinea as a dietary staple of African elites, their military retinues, prosperous free persons, and the resident Europeans who accepted the grain as a partial substitute for wheat flour. Wilhelm Müller, a Lutheran pastor who served at the Danish Fredericksburg fort along the Gold Coast from 1662 to 1669, mentioned a distinct European preference for bread made from millet over sorghum.[22] The peasantry grew millet in dry areas and sorghum, yams, and plantains in regions of higher rainfall as a subsistence preference and as tribute to their overlords.[23]

Agricultural slaves, on the other hand, were put to work in local fields producing crops that matured quickly or were adapted to specific ecological conditions. In semiarid environments, they grew millet and sorghum; in wetter regions, plantains, yams, and increasingly, maize. One slave trader operating in the Niger Delta reported slaves' subsistence preference for yams. He claimed that "no other food will keep them," with maize and other imported food "disagreeing with their stomachs."[24] Recognition of such preferences may have influenced what Europeans purchased as food for the Atlantic crossing.

Nevertheless, maize became a crop increasingly emblematic of slave labor and sustenance. It was produced by agricultural slaves and those awaiting deportation. It fed the growing segment of the African population remanded to coastal port settlements and destined for American plantations. Surplus production was routinely sold to provision slave ships.[25]

The cultivation of maize along the Lower Guinea Coast came largely at the expense of sorghum. Fields traditionally planted to sorghum were increasingly taken over by maize. While maize could not compete in the ecological conditions that favored pearl millet—sandy soils, low rainfall, and aridity—it grew in the humid clay soils where sorghum was traditionally planted. Maize's decisive advantage in food-intensification strategies was its yield. Productivity typically doubled that of sorghum. Maturing in as few as three months after sowing gave it a great advantage over the four to six months needed to harvest sorghum. In a single growing season two crops of maize could be effectively planted to one of sorghum, representing a fourfold improvement in output.[26] The demand of the Atlantic slave trade for surplus food favored the faster-growing maize and millet.

The land reserved for sorghum cultivation along the Lower Guinea littoral contracted to the acreage needed for feed and for brewing beer. Willem Bosman, a resident Dutch slave trader on the Gold Coast at the end of the seventeenth century, described how sorghum (identified by its red grains) contributed to the brewing process, but now included maize ("Great Milhio"):

> Here is also a third Sort of *Milhio,* like the last [millet]. . . . Its Grain is reddish, and must continue in the Ground seven or eight Months before it is full ripe. This is not eaten, but mixed with the *Great Milhio,* to brew withal, the Negros believing it strengthens the Beer. . . . The Negro Women are well skilled in brewing Beer. . . . All People here, the Slaves not excepted, drink only Beer; for their Water being . . . consequently very unwholsome in this hot Country; or drinking it but a few Days only, unavoidably brings-on a Fever.[27]

Bosman's description is typical of European accounts during this period, which suffer from a confusing and casual interchange of names. "Millet" is indiscriminately employed for both pearl millet (*Pennisetum* spp.) and sorghum (*Sorghum* spp.). Portuguese references compound the problem: *milho* was applied to both maize and sorghum, perhaps due to the similarity of the plants in their vegetative cycle. Only when the plants mature are the two easily distinguished, since sorghum produces a tassel of exposed grains, whereas maize has an enveloping husk. Sorghum grains are larger than those of millet, which cover a head that resembles a cattail. It is not always possible to determine whether sorghum, millet, or maize is meant in European accounts. All three were grown in Africa and sold as provisions to slave ships. Nonetheless, the subsistence transformations occurring along the Lower Guinea Coast draw our attention to important ecological factors that provide clues to the true identity of the plant referenced.[28]

But maize enjoyed still other advantages. It can be eaten fresh on the cob or stored as a dried grain. With its protective husk, maize better resists predation by birds and insects. Sorghum and millet, with their exposed grains, require more diligence to keep birds off the ripening heads, and their hulls must be removed with a mortar and pestle.[29]

African food surpluses were vital to the functioning of the commerce in human beings. The deepening of the Atlantic slave trade along the Lower Guinea Coast increased the demand for African-grown food, while reducing dietary choices for the enslaved. The insatiable European appetite for slaves was removing the continent's youthful laborers at the same historical moment that it was accelerating demand for surplus food to provision them. African elites responded by intensifying agricultural production through a variety of coercive relationships and by adopting new crops, such as maize and manioc.

In regions too arid for the cultivation of maize, the reorganization of agriculture for the transatlantic slave trade depended on increasing production of the traditional drought-tolerant cereals, sorghum and millet. In the midst of ever-present threats of drought and famine, raiding militias demanded surplus food, as did slave-ship captains seeking provisions. Amid this pressure to produce more, the rights of farmers eroded, as many communities were placed in increasingly onerous labor relations by their military protectors or were made indigenous slaves to work in agriculture. As demand for cereals rose, so did the need for female labor to cultivate and prepare them for consumption. In this context of social and political instability, new conventions evolved that influenced the master-slave relationship, gender norms, and land access for subsistence security.

AFRICAN CEREALS IN THE INTENSIFICATION OF FOOD PRODUCTION ALONG THE SENEGAL RIVER

For European ships journeying south of the Sahara, the Senegal River marked the transition to agricultural societies of the Upper Guinea Coast. The Senegal flows through an arid landscape into the African interior, where drought-tolerant millet and sorghum have long been the principal foodstaples. A single abbreviated wet season (from June/July to October) has since ancient times made cereal agriculture and animal husbandry pivotal to human settlement in the region. Pearl millet is traditionally planted on sandy soils with rainfall, sorghum on moisture-holding floodplains. But despite the cultivation of cereals suitable for low rainfall conditions and the utilization of two distinctive environments for food production, drought has long posed a recurring threat to human survival. During the centuries of the Atlantic slave trade, the number of people dependent on this vulnerable agricultural system dramatically increased. When cyclical droughts occurred, famine often followed, and as the land's ability to support people collapsed, more were cast into Atlantic slavery. Drought swelled the cargo holds of slave ships bound for the New World.

The development of the slave trade along the Senegal River fueled a demand for surplus food. As in Lower Guinea, slavery depended upon the exertions of agricultural slaves and peasants from whom predatory militaries exacted surplus production as tribute. The Senegal is navigable for hundreds of miles upstream, which provided slave raiders access to the African interior, where rural peoples could be attacked and captured. Once taken, slaves endured a march to the coast in human caravans (coffles) or a voyage downstream by boat.

During the four months that captives typically spent on the coast awaiting deportation, they were fed with locally produced grain. Town dwellers of the coastal ports also depended on local foodstuffs for their sustenance. Moreover, arriving slave ships carrying only partial stores expected to replenish them in order to provision their captives. James Searing estimates the annual cereal demand of the Atlantic trade from ports of the lower Senegal River at 950 tons in the eighteenth century.[30] In the Senegal River valley, a region where maize did not flourish, provisions for the slave trade depended vitally upon surplus production of pearl millet and sorghum.

Millet and sorghum cannot be consumed unless first milled to remove their indigestible husks. During the period of the transatlantic slave trade, the only way to carry out this critical step was by hand, with mortar and pestle. As milling and food preparation in Africa were traditionally women's work, the provision trade relied on the specialized skills of enslaved women to clean the grain and transform it into flour or porridge. This increased the need for female labor. Many slave-raiding expeditions relied on women to prepare and cook food.[31] In fact, grain pounders were known along the Senegal as *pileuses,* in recognition of the critical role of females in milling cereals. Enslaved women formed essential crewmembers of slave-raiding convoys along the Senegal River. They were brought along to mill grain and prepare food for soldiers and captives. One eighteenth-century French observer of slavers operating along the Senegal reported that an ideal barge crew of thirty-five would include four *pileuses* "to pound the millet and dress the victuals" for slaving missions.[32] The trade in grains depended vitally on female labor to transform cereals into food. Reliance upon enslaved women to process millet was not, however, confined to these fluvial voyages along the Senegal. It also at times occurred during the Atlantic crossing, on slave ships provisioned with African foodstuffs.

African women prepared a variety of food products from millet and sorghum. Jean Barbot described some of them from observations made during two slaving voyages (1678–79, 1681–82):

> Now I shall tell you of what that work consists, the first part being the pounding and preparing of millet. These women are also responsible for preparing food, as I told you before. They make *couscou*[s] in the following way (this being their staple food, taking the place of bread). They take some of the millet flour which is stored in large reed or straw baskets, and pass it through a sieve to separate out the bran and any dirt which may have got

into it. . . . When the millet is cooked . . . they also make some small cakes, cooking them on stones until really hot. *Couscou*[s] is the food of the better off. The peasants and the poor make do with another kind of cooked millet called *sanglet*. It is hardly more than millet bran. . . . This is what they normally feed their slaves on.[33]

Barbot's description indicates that slaves were given a rudimentary preparation of millet that was less laborious to prepare. Daily sustenance consisted of a thick porridge of millet and sorghum flour rather than couscous. The preparation of couscous involved considerably more labor and was reserved for the privileged. Valentim Fernandes, writing circa 1506–10, summarized the reports of Portuguese explorers compiled over the previous half century that describe couscous as the principal dish of Senegal.[34] In North Africa and Muslim Iberia, semolina wheat substituted for the pearl millet with which couscous was traditionally made in the Sahel. Its preparation began by pounding the grain to a coarsely ground flour, which was sprinkled with water and rolled between the fingers to form small granules. These were next passed through a sieve, dried, and prior to consumption, steamed. While enslavement along the Senegal brought little change in the amount of millet and sorghum traditionally consumed as a foodstaple, the coarser formulations that trade slaves received became, as elsewhere along the Guinea coast, the centerpiece of an unvarying and monotonous diet.

The importance of female slaves in the processing and preparation of millet and sorghum for food influenced the gender dynamics of the regional slave trade. Their crucial role in transforming surplus cereal production into food may explain why enslaved women were disproportionately retained in Senegambia during the Atlantic slave trade, which exported almost twice as many men as women to New World plantation economies.[35]

The production of surplus food took place under different conditions of coercion. Slaves taken in the interior could only be shipped downstream during the three months of high river water, and if they arrived at the coast during the rainy season when slave ships were less frequent, they were often made to labor in food production, growing millet and producing other valuable commodities, such as cloth, which could be exchanged for grain. The practice of using slaves as temporary agricultural workers was already in place in the mid-fifteenth century, when Alvise da Cadamosto, a crewmember on a Portuguese caravel that traveled to the Senegal River, reported a Wolof king turning enslaved war captives to agricultural production. In this manner, trade slaves were made to work on behalf of their own maintenance.[36]

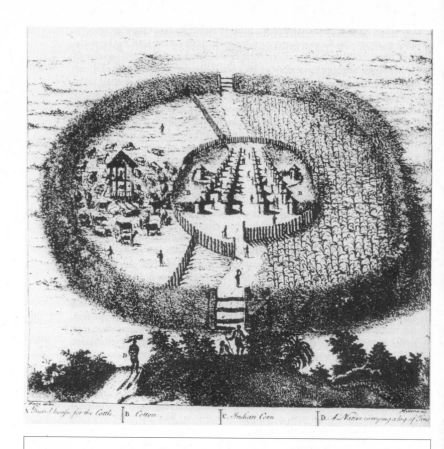

Legend:

A. Guard house for the cattle
B. Cotton field

C. Indian corn (maize)
D. "Native" carrying firewood

FIGURE 3.6.   *Fula Village and Plantation,* near Gambia River, by Francis
Moore, early 1730s.

SOURCE: Moore, *Travels into the Inland Parts of Africa,* opposite p. 35.

Peasant communities also produced surplus food (an example from the
cattle-keeping Fula people in Gambia is shown in figure 3.6). The turmoil induced by slave raiders forced vulnerable rural communities to submit to extor-
tionate relationships imposed by dominant elites. In exchange for military protection, peasants paid tribute in surplus grain, cattle, other foodstuffs, and local
products. This process weakened the position of independent agricultural producers in Senegambia as it had along the Gold Coast, forcing them to surrender

a significant portion of what they produced. Despite the promises of protection, they nonetheless lived under the constant threat of enslavement, as the military defeat of their guardians could easily lead to their capture by the victors.[37]

Africans held in indigenous slavery grew a considerable amount of the surplus grain produced for coastal settlements and slave ships. At the end of the eighteenth century, which delivered more Africans into bondage than any previous century, explorer Mungo Park noted that slaves performed most of the agricultural work in Senegambia.[38] They worked in a variety of labor systems. The newly enslaved produced grain in labor groups under close surveillance. They held no right to the surplus they produced and remained dependent on their owners for subsistence. Usually ill fed, they experienced little diversity in their daily diet. At the other extreme of supervision were slaves settled in agricultural villages, known throughout Senegambia as *rumbdés*. They produced surplus food for African elites and looked after their own subsistence.[39] In a somewhat idealized portrait, the French explorer Gaspard Mollien provides an understanding of some of the serflike rights these satellite slave villages held: "In every village, several inhabitants assemble their slaves, and make them build themselves huts close to each other; this is called a rumbdé.... These slaves, at least, nominally so, cultivated the plantations of their masters, and accompany them for carrying their burdens when they travel. They are never sold in old age, or when born in the country. Any departure from this practice would cause the desertion of the whole rumbdé, but the slave who misbehaves, is delivered. up to the master to be sold."[40]

In between the categories of newly captured agricultural slaves and those settled in *rumbdés* was another form of indigenous slavery. This included the acculturated slaves who had gained the confidence of their masters or who were born into a trusted slave family. These slaves achieved the right to a plot of land for their own cultivation. Norms regulated which days of the week and time of the day belonged to the master.[41] The convention of granting some slaves a food plot and time to work it effectively put them in charge of their own subsistence. Importantly, it gave them the opportunity to grow and consume the food they preferred.

This convention already existed in Senegambia during the earliest period of the Atlantic slave trade. In the early 1500s, Valentim Fernandes reported in Senegambia that "the slaves of this country work and earn for their master during six days and the seventh day they earn what they need to live the other six." Searing observes that after a workday for the master that typically lasted from sunrise until two or three in the afternoon, Wolof slaves worked their own subsistence plots. They were also exempted from labor for their masters one or two days a week.[42] Similar practices were also noted in the early sixteenth century on São Tomé, where agricultural slaves retained rights that specified days of the week or hours of the day for working their own autonomous food plots.[43] The rights of indigenous slaves distinguished them from the enslaved Africans traded to Europeans as chattel, who had no rights whatsoever. It provides context for understanding the comments of André Alvares de Almada, who wrote in 1594 of "Fula slaves ruling the Wolofs."[44] Such rights held by indigenous slaves were not available to trade slaves.

Access to land enabled some people placed in indigenous slavery to establish normative limits on the hours and days they worked for their masters in exchange for providing their masters surplus grain. Independent production gave slaves the means to improve their diet, to reassert subsistence preferences, and to diversify daily food intake. It provided the potential of supporting a family, of softening the psychological blow of having been exiled from one's community and kin. The convention made it possible to recreate kinship, culture, and identity and to establish some measure of security. This was an important hedge in a world where sale as human chattel was a common occurrence and drought-induced food shortages an added vulnerability. Access to an independent plot narrowed the divide that separated a slave from a free person.[45]

In the formative period of New World plantation societies, slaves would seek to renegotiate the convention of independent food production, but with varying degrees of success. However, the struggle of slaves to achieve it underscores the significance of subsistence to displacement, survival, identity, and memory.

FOUR

African Food and the Atlantic Crossing

The Blacks eat large quantities of these fruits . . . and are used to putting them
in water in order to make it more tasty.

JOÃO ANTÓNIO CAVAZZI DE MONTECÚCCOLO

Angola, ca. 1660

There is also a fruit called "cola" . . . which quenches the thirst and makes water
delicious to those who make use of it.

JEAN BARBOT

ca. 1688

THE LITERATURE ON THE TRANSATLANTIC slave trade largely focuses on
the commodities exchanged between Europe, Africa, and the Americas. In the
classic formulation known as the Triangle Trade, Europeans sold firearms, iron
bars, spirits, textiles, and beads in Africa in order to purchase slaves. Slaves were
carried from there to the New World, and from the Americas the commodities
produced by enslaved labor were transported to Europe. Sugar, coffee, and tobacco became items of everyday European consumption. But missing from this
narrative is the critical importance of food to the entire enterprise. Only
through the availability of food could each segment of the Triangle Trade op-
erate. The exchange of commodities between three Atlantic continents depended crucially on keeping the enslaved Africans that were its key component
alive over the Atlantic crossing. The foodstaples that provisioned slave ships provide the bridge for understanding the means by which African dietary staples
arrived in New World plantation food fields.

Although maize, manioc, and other Amerindian crops helped to fulfill the
enormous subsistence demand created by the transatlantic traffic in human
beings, African species were likely put aboard every single ship that crossed the
Middle Passage. They were stowed as provisions, meat, medicines, spices, lamp
oil—even flavorings to improve drinking-water quality. Slave ships became the

unwitting vessels of Africa's botanical heritage by carrying seeds, tubers, and the people who valued them to the Americas. Similar to the way in which European botanists of the time used the African bottleneck gourd as a watertight container for shipping specimens to scientific and commercial collaborators, the slave ship likewise conveyed African species to the Americas.

This botanical legacy is not evident in the commodity focus of scholarship on the Atlantic slave trade and plantation economies. It is better illuminated through the lens of subsistence, the food that served as the daily fare of the enslaved. The African components of the Columbian Exchange come into sharp relief by engaging how the need for food shaped the transatlantic slave trade.

The captains of slave ships who purchased African foodstaples knew little of the crops with which they provisioned their vessels, much less the ways these unfamiliar foods were grown. In fact, slavers often referred to these tropical African staples by generic terms: cereals were known as the "corn" of Guinea, beans by the slave ports where they were loaded, and all manner of tubers were simply called "yams." Slavers' fundamental concern was an expeditious provisioning and speedy departure so that they could safely—and profitably—land a boatload of captives on the other side of the Atlantic.

Although the African components of the Columbian Exchange journeyed to the Americas because of slavers' actions, the establishment of African foods in plantation societies followed a different course. The plant introductions owed their presence in slave food fields to Africans themselves, who took the initiative in planting their dietary preferences from the leftover provisions that at times fortuitously remained from slave voyages. This testifies to an unusual form of botanical transfer, one that occurred with the arrival of enslaved Africans and was driven by their resilience in adversity.

AFRICAN CROPS IN SLAVE SHIP PROVISIONS

Perhaps as many as twelve million people were forcibly deported from Africa over three and a half centuries of Atlantic slavery. Of this number, at least ten million made it to the New World alive. The forced migration of enslaved Africans to the Americas involved an almost inconceivable number of transatlantic journeys. There is now supporting documentation for at least thirty-five thousand slave voyages to the Americas, but many without doubt went unrecorded. More than seventeen thousand occurred in the eighteenth century alone.[1] Even this number puts into perspective the enormity of the demand for slave-ship provisions along Africa's Atlantic coast. For every ship that boarded slaves in Africa, the success of the enterprise rested on the ability to find enough food surpluses to keep alive a boatload of human beings—often several hundred—on the open ocean for the duration of the Atlantic crossing.

Ship captains estimated the subsistence needs of their captives from a formula that allowed two daily meals on a transatlantic voyage ideally expected to last between three and six weeks. The French slave ship *The Diligent*, for instance, used a generally accepted provisioning rule of one ton of foodstuffs per ten captives.[2] Some estimates of food purchases are evident in the documentary record. Captain Thomas Phillips purchased five tons of rice along the Rice Coast on his 1693–94 voyage; James Barbot bought rice in the same region for his slave voyage across the Atlantic in 1699. One ship captain, John Matthews, estimated that he needed 700 to 1,000 tons of the grain to feed the 3,000 to 3,500 slaves awaiting shipment from Sierra Leone.[3] For the 250 slaves the *Sandown* carried to Jamaica in 1793, Samuel Gamble purchased more than eight tons of rice, cleaned as well as in the husk (Gamble purchased the grain in the husk because it was offered at a lower price than the milled rice brought from the interior by caravan).[4] In seventeenth-century Angola, slave-ship captains calculated a measure of just under two liters of manioc flour per captive per day, plus one-fifth of a liter of African beans or "corn," or flour made from the shell of oil palm nuts.[5] In 1750 John Newton loaded his ship with cowpeas and nearly eight tons of rice for the 200 slaves he carried across the Middle Passage. The slave ships *Fredensborg* and *The Diligent* stocked millet (likely sorghum) and cowpeas as provisions; others carried pigeon peas. In the Bight of Benin, yams and plantains frequently were sold as provision. *The Diligent* purchased at Príncipe one thousand plantains as food for its captives.[6]

A slave ship departing Europe for the African coast brought some food stores, such as salted meat and fish, cheese, biscuits and wheat flour, beer and wine, intended for the most part for the officers and crew. After months at sea it became a monotonous and often moldering diet, which was relieved by additional food purchases once the ship arrived in Guinea. A variety of African foodstaples and essential goods could be bought from African societies, including grains, legumes, even supplies of hippopotamus and elephant meat.[7] In one transaction, slave trader Theophilus Conneau described the diverse items offered for purchase. A seven-hundred member Fula caravan arrived at the Atlantic coast with thirty-six bullocks for sale, an unspecified number of live sheep and goats, fifteen tons of rice along with forty slaves, thirty-five hundred cattle hides, and nine hundred pounds of beeswax.[8]

However, this critical provisioning could take weeks, if not months, depending on circumstances such as food availability, competition from other embarking slave ships, and the number of slaves to feed. These factors could compel a slave ship to put into many different ports along the African coast in order to fill its stores. For instance, William Littleton, who traded in the drought-prone Senegambia region for eleven years in the 1760s and 1770s, observed that slave ships calling there seldom obtained sufficient quantity of millet and sorghum for the Middle Passage, forcing them to journey south to the Grain Coast for (African) rice.[9] Areas known for their ability to supply slaves often could not provide adequate provisions to all the ships that called there. The sheer number of slave ships operating along the Slave Coast meant that most had to travel elsewhere to find provisions. Like the eponymous Banana (Plantain) Islands, São Tomé and Príncipe became specialized as provisioning way stations that grew plantains and other tropical tubers.[10]

While captains hoped to fill their larders quickly, sometimes several months went by before they could secure the necessary supplies. Delays greatly increased the risk of disease outbreak as well as the threat of revolt by slaves already boarded. Some accounts of slave traders mention the scarcity of provisions, rather than the availability of slaves, as the main impediment to embarkation. One seventeenth-century trader along the Slave Coast addressed the issue of seasonal food shortages:

> In the months of August and September, a man may get in his compliment of slaves much sooner than he can have the necessary quantity of yams, to subsist them. But a ship loading slaves there in January, February, etc. when yams are very plentiful, the first thing to be done, is to take them in, and afterwards the slaves. A ship that takes in five hundred slaves, must provide above a hundred thousand yams; which is very difficult, because it is hard to stow them, by reason they take up so much room; and yet no less ought to be provided, the slaves there being of such a constitution, that no other food will keep them; Indian corn, [fava] beans, and Mandioca [manioc] disagreeing with their stomach; so that they sicken and die apace.[11]

But the problem of food availability was a continuing one. In the early nineteenth century, the Portuguese governor of Angola ordered arriving slave ships to bring sufficient manioc flour for the return voyage to Brazil.[12]

Stimulating demand for traditional African foodstuffs was the view that more Africans survived slave voyages if they were fed the food to which they were accustomed. The belief that familiar foods reduced slave mortality influenced slaver purchasing preferences, though they were not always realized. But the perceived linkage between customary subsistence staples and survival contributed to a sustained demand by slavers for traditional African foods.[13]

However, provisioning a slave ship with African staples hardly meant that slaves were being served customary food. Their meals typically consisted of beans mixed with some sort of starch. Legumes formed an important component of the food slaves were fed. Europeans at times substituted horse (fava) beans for the indigenous ones they found along the Guinea coast: black-eyed peas, pigeon peas, lablab or hyacinth beans, and the indigenous Bambara groundnut.[14] These were mixed with cereals, yams, or the flour of manioc into a starchy gruel, made only slightly more palatable by seasoning with African palm oil and melegueta pepper.[15]

WATER AND THE MIDDLE PASSAGE

When European slave ships arrived along the African coast—several weeks journey from northern Europe to the Cape Verde Islands, an additional month to the Gold Coast, some three to four weeks in total to Angola—they needed to replenish dwindling, and often stale, supplies of water as well as firewood for cooking fuel.[16] Ship captains paid a fee or provided gifts to African chieftains for the right to take on water and cut trees in places where they were known to be available. The success of any slave voyage depended vitally on the availability of drinking water and food. Inadequate supplies could result in high mortality. Each vessel required below deck a considerable storage space for holding barrels of water, which could not be replenished en route.

Captains estimated their water needs based on anticipated days at sea and daily per person consumption. The log from *The Diligent*, which left France for Africa's Slave Coast in 1731, calculated one sixty-gallon cask of water per captive. This would provide an allotment of three quarts daily per slave on a typical transatlantic voyage of eighty days, of which one quart would be used in cooking the souplike gruel they were fed. However, half that amount and less was observed as the daily ration on many English, Danish, Dutch, and Portuguese slave voyages that took place over the eighteenth century. Because of the prevalence of dysentery aboard slave ships, some captains believed it best to provide slaves little water.[17] When an Atlantic crossing was prolonged because of calm seas, the amount of water allotted captives could be reduced to pitiful amounts, just spoonfuls with each meal.

Consumption of foul water was believed responsible for the dysentery and disease outbreaks that frequently occurred on slave ships. Captains were especially concerned about the quality of freshwater along the Slave Coast, which they believed caused worms and scurvy.[18] The search for supplies of good water led them to offshore islands or other specifically recommended locales along the coast:

> Tho' this [water] of St. Tome [São Tomé] keeps pretty well in casks, after it has once stunk, and is recovered. I would advise such as resort thither to victual their ships, to water in other places in the island, or in the middle of the town, through which the river runs, tho' it will cost double the labour and charges. For it is so essential a point, that the water taken aboard in slave-ships should be of the very best and cleanly, that it often contributes very much to save or destroy whole cargoes of them, according as it is good or bad; and rather than to run a risque, I would advise them to go to cape Lope [Gabon], Prince's Island [Príncipe], or Annobon for it; because many ships have lost the best part of their compliment of slaves by that water, in their passage from thence to America.[19]

On lengthy sea voyages a ship's water often spoiled: "In fact, it seems to have gone through cycles of going bad and curing itself again." Nonetheless, in the eyes of some slave dealers, this water, whether good or bad, was preferable be-cause it remained the substance to which Africans were naturally habituated. Some Dutch accounts mention that when slave vessels reached American ports, the captives were only gradually weaned from the stagnant water carried from Africa.[20]

Water stored in casks on long voyages was vulnerable to microbial growth, often unhealthy to drink, and frequently unpleasant to taste. Several additives were thought to act as water purifiers and to improve its palatability. One was brandy, which was also esteemed for its medicinal properties. Another was the African tamarind. One slave owner described how "in *Africa* the Negros make a Drink of it, mixed with Sugar, or Honey, and Water. They also preserve it as confection to cook and quench Thirst; and the Leaves chewed produce the same Effect." Slavers adopted the African practice of making the seeds into a tart pulp to improve the flavor of water. They also stocked the seeds as a medicinal, be-lieving that "tamarinds are excellent for combatting scurvy."[21]

Another African plant that played an especially important role in improving the taste of fetid water was the African kola nut. When chewed without swal-lowing, it made, in the words of the French slaving captain Chevalier Des Marchais, who traveled to western Africa between 1725 and 1727, the "bitter-est, or sourest Things taste Sweet after it."[22] Visiting the Gambia River in 1623, Richard Jobson described them "like the bigger Sort of Chestnuts, flat on both

Sides, but the Shell is not hard. The Taste is bitter, but the Effect is so esteemed, that ten of them is a Present for a King; for the very River-Water, drank after chewing it, relishes like White-Wine, and as if mixed with Sugar." One European, Wilhelm Johann Müller, based at Danish Fort Frederiksborg on the Gold Coast (1662–69), mentioned Africans using the kola nut "when drinking so that their drink may taste better," while others wrote that it imparted a pleasant, sweet taste to water and forestalled hunger pangs.[23]

Willem Bosman, a Dutch trader who worked on the Gold Coast for fourteen years at the end of the seventeenth century, mentioned that "not only the *Negros*, but some of the *Europeans* are infatuated to this fruit" for the kola nut's value as a diuretic and to "relish" palm wine (which otherwise became sour after a day or so).[24] Jean Barbot, who participated in two slaving voyages to Guinea in the late seventeenth century, elaborated on a plant that he depicted and likely employed: "There is also a fruit called 'cola' and by others, 'cocters', which quenches the thirst and makes water delicious to those who make use of it. It is a kind of chestnut, with a bitter taste. . . . Here is a drawing [figure 4.1], showing it both whole and cut open down the middle. I give it natural size. The outside is red mixed with blue and the inside violet and brown."[25]

Africans prized kola as a stimulant; its level of caffeine considerably exceeds that of coffee (Africa's other notable indigenous stimulant). Over the sixteenth century, ancient kola nut trade networks expanded to burgeoning coastal markets, perhaps in part because the nuts were loaded aboard slave ships to curb hunger and thirst and to freshen the taste of stagnant water and food. Recent scientific experiments show that stale or impure water becomes quite palatable after chewing kola. This may result from chemical changes affecting the tongue, which create an illusion of sweetness, or may be related to kola's high caffeine content.[26] These properties are succinctly captured in the advertising slogan of the world's favorite soft drink, "the pause that refreshes," whose signature ingredients are based in part upon the African plant.

Slave voyages undoubtedly contributed to the early presence of the kola nut in New World plantation societies. One report from 1634 mentions that newly landed African slaves in Cartagena were given kola nuts as a stimulant and medicinal.[27] Kola was certainly used for more than curing bad water. As a stimulant, kola may have already become an important medicinal component of Afro-Caribbean populations, as it was in Africa. Moreover, kola nuts are prominently featured in the liturgical practices of the Afro-Brazilian candomblé religion. The transport of the kola nut on slave ships provided the means to reestablish an important African stimulant in tropical America. The continued use of African language names for kola—*bissy, goora, obi*—that persist in tropical America at-

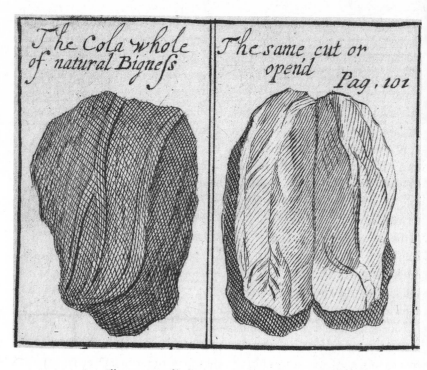

FIGURE 4.1.   Illustration of kola nut, near Cape Mount, Liberia, late
seventeenth century.

SOURCE: Barbot, "Description of the Coasts of North and South Guinea," in Churchill, *Collection
of Voyages and Travels*, vol. 5. pl. V, opposite p. 107.

test to the plant's enduring importance among populations descended from
enslaved Africans.[28]

FOOD PREPARATION AND
AFRICAN WOMEN ON SLAVE SHIPS

The proportion of male to female captives was often a concern of slave-ship captains, even if their preference was to buy adolescents and young adults. Drawing upon three slaving voyages between 1727 and 1730, the British captain William Snelgrave recommended a gender ratio of three males for every female captive; his French contemporary Chevalier Des Marchais, who also slaved in Lower Guinea, wrote in the 1720s that women should not form more than one-third of the ship's captives.[29] The recommended ratios were at best ideals. The gender composition of the captives on slave ships, while manifestly calculated with the perceived demands of foreign slave markets in mind, may not have been wholly determined by commercial or even arbitrary motives. It may also have been informed by recognition of a logistical need to keep a critical proportion of women onboard during the Middle Passage for the essential function of food preparation.

There were marked differences in the gender makeup of Africans available to slave merchants along Atlantic Africa. For instance, some regions of Upper Guinea, such as Senegambia, showed a distinct preference to retain women in indigenous slavery, which encouraged the export of twice as many males as females.[30] In fact, some slave captains in the area complained of shortages of women slaves. John Tozer wrote in 1704 from Gambia that his ship had "never caryed above 60 Men & all ye Rest woomen & children [elsewhere, but] here are all men." In the same year, Captain Weaver writing from Portodally, Senegal, experienced a similar problem and worried about the surfeit of men and its impact: "he hath not been able to purchase a woman and there must be at least 12 Women to dress the Victuals."[31]

Weaver's comment reveals his dependence on African women to assist in the preparation of food for captives during the Atlantic crossing. Pressing enslaved females into mess duty may have been common practice; it certainly appears to have been so on voyages from Senegambia. The number of slaves on board a slave ship could be in the hundreds, and the vastly outnumbered crew would be primarily occupied with nautical duties and guarding their human cargo. According to historian Hugh Thomas, "Female slaves were often asked to work the corn mill, the corn being put, perhaps with rice or peppers, into the bean soup." A journal entry from the slave ship *Mary* (outbound from Senegal), dated June 19, 1796, recorded female slaves milling rice and grinding another cereal, likely millet: "Men [crew] Emp[loye]d tending Slaves and Sundry Necessaries Jobs about the Ship. . . . The Women Cleaning Rice and Grinding Corn for corn cakes."[32]

From the start, food preparation influenced the basic design and outfitting of a slave ship. First, the wood-burning cooking area, normally located below deck, was removed to a position above deck. This was done to heighten security of the galley area in the event of a slave revolt. Slaver captains typically relocated the cast-iron cooking hearth to the quarterdeck near the lodgings of the ship's officers, where enslaved women and young children were held.[33] A nineteenth-century French lithograph illustrates the enclosed cooking area (with chimney) and its location in proximity to enslaved females (plate 2). Also shown is the wooden partition or barricade that segregated the male from female slaves on the ship.[34] A door in the partition facilitated the passage of food at mealtimes, when small groups of enslaved males were allowed on deck to eat. The lithograph shows a cauldron of food being passed through the barricade to male captives. Weapons and violence are prominent and pervasive—weapons to maintain control of a slave population that could outnumber the crew by a ratio of thirty or forty to one—but so is food, upon which a successful Atlantic crossing also depended.[35] The stove's location on the female side of the partition placed the women in easy proximity if drafted into food preparation for the ship's complement of captives.

Slave merchants sometimes bought food already processed, such as milled rice or couscous made from millet. Accounts of Portuguese traders on the Upper Guinea Coast from 1613 to 1614 even record purchases of the small-grained native African cereal fonio *(Digitaria exilis).*[36] But captains of slave ships also bought as supplies indigenous African cereals in the husk, that is to say, not yet milled. Rice, millet, and sorghum could be purchased more cheaply in this manner, but labor availability also may have affected how these grains were bought.[37] Loaded onto a slave ship, raw grain could not be eaten unless the husks were first removed, and this could only be done by hand. Milling, and food preparation in Africa in general, was traditionally the work of women.[38] The crews of slave ships (such as the *Mary*) that used women for "Cleaning Rice and Grinding Corn" simply exploited existing cultural practices by drafting enslaved females to pound and winnow unhusked African cereals. Captains Tozer and Weaver, for example, plied the region of the Upper Guinea Coast where the food supplies primarily consisted of millet, sorghum, and rice. If the cereals were not bought already milled, they would have to be processed onboard—or "dressed," as Weaver put it—perhaps illuminating his concern for the lack of female slaves. On such voyages, the specialized skills monopolized by African women may have been as prized as they were on the fluvial slaving expeditions along the Senegal River.

The milling of cereals was not easy work, as Richard Jobson underscored in his voyage to the Gambia River in 1620–21. "I am sure there is no woman can be under more servitude, with such great staves wee call Coole-Staves [pestles], beate and cleanse both the Rice, all manner of other graine they eate, which is onely womens worke, and very painefull."[39] Hand-milling was not only strenuous work; it also required skillful application of the African mortar and pestle. The African mortar, an upright hollowed-out cylinder carved from a tree trunk, and used with a handheld pestle, was in fact the only known way to separate rice from its husk in the Atlantic world until the advent of competent milling machines in the late eighteenth century.[40]

Slave ships that provisioned with grain still in the husk stocked milling devices appropriate for their stores: grinding stones or iron rollers for maize, the mortar and pestle for unhusked African cereals.[41] On a visit aboard an American slave ship that had stopped in Barbados en route to Savannah, Georgia, Dr. George Pinckard in early 1796 described slaves milling rice: "Their food is chiefly rice which they prepare by plain and simple boiling. . . . We saw several of them employed in beating the red husks off the rice, which was done by pounding the grain in wooden mortars, with wooden pestles, sufficiently long to allow them to stand upright while beating in mortars placed at their feet."[42]

This slave ship had arrived in Barbados from Africa's Grain Coast and evidently carried as provision rough or unmilled rice, which required pounding to ready it for consumption, even as the ship lay in port. The red husks in Pinckard's description moreover indicate that the rice was the indigenous African *glaberrima* species.

Other accounts of slave ships make the gendered basis of milling unhusked grain more explicit. The journal of the slave ship *Mary* described "females cleaning rice," while British naturalist Henry Smeathman, writing from Sierra Leone in the 1770s, observed: "Alas! What a scene of misery and distress is a full slaved ship in the rains. The clanking of chains, the groans of the sick and the stench of the whole is scarce supportable . . . two or three slaves thrown overboard every other day dying of fever, flux, measles, worms all together. All the day the chains rattling or the sound of the armourer rivetting some poor devil just arrived in galling heavy irons. The women slaves in one part beating rice in mortars to cleanse it for cooking."[43]

An eighteenth-century oil painting (ca. 1785) provides additional evidence that African women were indeed put to work hand-milling unhusked cereals aboard slave ships. Detail from a depiction of the Danish slave ship *Fredensborg* shows two female slaves at work on the quarterdeck near the mizzenmast (plate 3). Each one holds a lifted pestle to pound grain in the mortar between them. The *Fredensborg* was known to be provisioned with "millet," and the women are likely milling grain that was purchased in the husk.[44] The painting provides visual confirmation that enslaved women labored in food preparation on some slave ships and that they used the African mortar-and-pestle technology for milling.

While it was quite common for slaving voyages to end with depleted stores and their human cargo nearly starved, this was not always the case. Some ships reached the Americas with leftover provisions. Pinckard's comments on the slave ship he visited in Barbados provide one instance. Another account from South Carolina in the 1690s traces one early rice shipment to the arrival of a "*Portuguese* vessel ... with slaves from the east, with a considerable quantity of rice, being the ship's provision ... but was not sufficient to supply the demand of all those that would have procured it to plant."[45] It is important to note that rice suitable for planting must be unhusked; that is, unmilled grains with the outer hulls still intact. Once the husk is removed by milling, the grain cannot be planted and is useful only as food. Taken together, these historical observations directly implicate the slave ship as an agent of botanical dispersal. Each example captures this discrete and seemingly accidental mechanism of seed transfer from Africa. Thus, we find in the ports of New World plantation societies a propitious alignment of the fundamental ingredients for establishing African cereals in the Americas: seeds of African provenance, the presence of people skilled in their cultivation and processing, and individuals for whom the cereals were a preferred dietary staple.

The slave ship is also the vessel for rice introduction in the oral histories of the descendants of runaway slaves (known as Maroons). These stories, which can be heard even today in isolated Maroon communities across northeastern South America, begin with the arrival of a slave ship. In a version from French Guiana, an African woman onboard the ship takes rice seeds and hides them in her hair before she is disembarked. This, her Maroon descendants claim, is how they came to grow rice.[46] Another version of the same narrative, from Pará, Brazil, makes the woman's children the agents of rice dissemination. The mother, afraid that she cannot prevent their imminent sale and separation, tucks grains in her children's hair, bestowing a gift of lifesaving food from Africa. Like other disembarkation narratives, this version underscores the role of a female ancestor in promoting the diffusion of an African dietary preference in the Americas.[47]

However, a story from neighboring Maranhão adds an illuminating coda to the Pará narrative. The Itapecurú-Mearim watershed, located inland of São Luis, is a low-lying area where rice initially was planted as a subsistence staple before becoming, in the mid-eighteenth century, the regional locus of a plantation economy.[48] This version makes explicit the African origins of rice and undercuts claims of European agency:

> An enslaved African woman, unable to prevent her children's sale into slavery, placed some rice seeds in their hair so they would be able to eat after the ship reached its destination. As their hair was very thick, she thought the grains would go undiscovered. However, the planter who bought them found the grains. In running his hands through one child's hair, he pulled out the seeds and demanded to know what they were. The child replied, "This is food from Africa." So this is the way rice came to Brazil, through the Africans, who smuggled the seeds in their hair.[49]

By embellishing the Pará tale in a crucial way, the Maranhão version illuminates another layer of social memory and historical consciousness. When the white man discovers the seeds in the child's hair, he reveals complete ignorance of them. Upon learning what they are, he seizes the grains as his own. The child's explanation of the significance of rice as African food constitutes the planter's first lesson in the uses of rice. In drawing attention to seeds and their appropriation by planters, the story symbolizes a transfer of knowledge from enslaved African to white owner. In this case, not only is the child the instructor of the man, the slave is the tutor of the master, and the black the teacher of the white. The usual power relations between master and slave are inverted, as are the traditional accounts of European inventiveness and ability. African seeds and funds of knowledge benefit the white man, and the story figuratively encapsulates the historical transformation of rice from an African-grown subsistence crop to a slave-produced plantation commodity.

Seen as allegory, the narrative challenges a Eurocentric narrative of agricultural development in the American tropics that would exclude African agency and initiative. The development of rice plantations in Maranhão depended on the appropriation of African expertise and labor, symbolized as the slave owner taking away the rice seeds brought from Africa. In casting the European "discovery" of rice as something akin to theft, the Maranhão story reclaims rice as a legacy of the African diaspora.

In attributing rice beginnings to their ancestors, Maroon legends reveal the ways in which the enslaved gave meaning to the traumatic experiences of their own past while remembering the role of rice in helping them resist bondage and survive as fugitives from plantation societies. These oral histories offer a counternarrative to the way transoceanic seed transfers are discussed in Columbian Exchange accounts. They substitute the usual agents of global seed dispersal—European navigators, colonists, and men of science—with enslaved women whose deliberate efforts to sequester rice grains helped reestablish an African foodstaple in plantation societies. The stories link plant transfers to the transatlantic slave trade, African initiative, and the dietary preferences of the enslaved. Each narrative sharply contrasts with written accounts that credit European mariners with bringing rice seed from Asia and the initiative and ingenuity of slaveholders in "discovering" the suitability of rice as a plantation commodity in the New World. Importantly, the vessels of European botanical transfers metamorphose into slave ships carrying African peoples and seeds.

These Maroon stories need not be taken literally. While they cannot be substantiated by European accounts, they do accord with certain documented episodes. Slave ships occasionally arrived in the Americas with leftover rice suit-

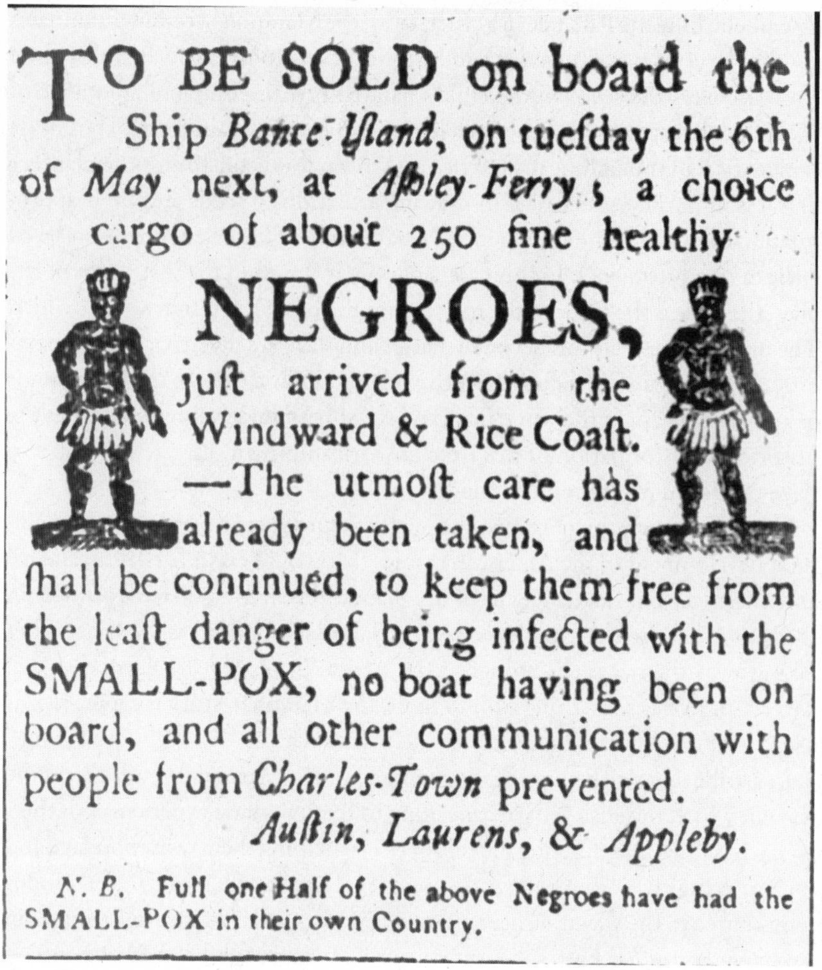

> TO BE SOLD, on board the Ship *Bance-Island*, on tuesday the 6th of *May* next, at *Ashley-Ferry*; a choice cargo of about 250 fine healthy NEGROES, just arrived from the Windward & Rice Coast. —The utmost care has already been taken, and shall be continued, to keep them free from the least danger of being infected with the SMALL-POX, no boat having been on board, and all other communication with people from *Charles-Town* prevented.
> *Austin, Laurens, & Appleby.*
>
> *N. B.* Full one Half of the above Negroes have had the SMALL-POX in their own Country.

FIGURE 4.2. Advertisement for slave sale, announcing the arrival in "Charles-Town" (Charleston) of Africans who specialized in rice cultivation, ca. 1760.
SOURCE: Courtesy of the Library of Congress, Prints and Photographs Division LC-USZ 62–10293.

able for planting, the technology for removing the hulls, and people for whom it was a dietary staple. We may never know exactly how African seeds and root-stock went from the ship to the shore, from the pier to the plot. But we know that they did, and that this occurred early in the colonial period and in many plantation societies.

An unintended, and perhaps ironic, consequence of slave voyages is that some may have indeed provided slaves the opportunity to access the seeds and roots of African crops as planting stock. In leftover provision on slave ships, enslaved Africans found the means to reestablish the continent's principal food crops in the Western Hemisphere. The slave voyage may have ended at the auction block (figure 4.2), but Africa's botanical legacy did not. African food plants abetted the struggle to keep alive both the body and the social memory of lost homes and lives. These plants nourished and sustained African lives and African identities in desperate and brutal circumstances.

FIVE

Maroon Subsistence Strategies

In slavery, there was hardly anything to eat. It was at the place called Providence Plantation. They whipped you . . . Then they would give you a bit of plain rice in a calabash. . . . And the gods told them that this is no way for human beings to live. They would help them. Let each person go where he could. So they ran.

A SARAMAKA MAROON

ca. 1970s

Perhaps the people of Brazil have to thank my great-great grandfather for these plants.

JOÃO RIBEIRO

quilombo community leader, 2005

THE FIRST GENERATIONS OF AFRICANS who arrived in the Americas found themselves in environments not yet wholly transformed by European colonization. The outposts of empire fronted vast tracts of unknown territory, and these provided refuge for many enslaved Africans who opted for freedom. Despite the dangers, some runaways were able to form or join free communities of other escapees in the hinterlands that lay outside the bounds of colonial authority. They escaped to rugged environments whose inaccessibility discouraged pursuit and provided defensible shelter. Indeed, through most of the eighteenth century, more slaves in the New World gained their freedom as escaped maroons than by legal manumission.[1] In the mountainous regions of the Caribbean and mainland tropical America, in concealing rainforests above river rapids in the Guianas and Colombia, in swamps and other inhospitable refugia of mainland North America and Mexico, communities of runaways persistently took root beyond the peripheries of white control. Survival depended on their ability to evade detection and capture, or to wage battle—but also on the skills and knowledge that could wrest reliable supplies of food from the harsh surroundings.

Wherever maroon communities gained a toehold, colonial authorities inevitably recruited armed militias to seek them out, destroy their redoubts, and capture the inhabitants. They often met fierce resistance. Palmares—the most famous maroon settlement of Brazil's early history—ably defended itself from repeated military attack for nearly the entire seventeenth century.[2] Otherwise, paramilitary forces managed to reduce most enclaves, but not all. In some areas the numbers of maroons were so large and defensive tactics so effective that the colonies negotiated treaties granting some groups their freedom, as occurred repeatedly in Dutch Guiana in the seventeenth and eighteenth centuries.[3] Other maroon communities evaded destruction by retreating to more clandestine areas until emancipation by the state made their freedom a fact. The Amazon Basin supported many such refugee communities, as did mountainous regions of Brazil, Hispaniola, Jamaica, Cuba, and the French Caribbean. Today, maroon societies continue to occupy some of these remote locales, and one can still discern a distinctive culture whose agricultural practices and plants reveal unmistakable links to Africa. Even vanquished maroon settlements speak to us through the accounts and maps left by the expeditions sent against them. From all of these sources, we learn of the fundamental importance of the agricultural systems that supported subsistence, and the crops that sustained the maroon struggle for freedom.

Agriculture shaped the lives of the majority of Africans landed in the Americas. Subsistence practices and food preferences of slaves offer one way to see manifestations of the African botanical legacy in the New World. Another is through the subsistence strategies developed by maroons. Liberated from the ceaseless toil of chattel slavery, maroons hunted and farmed for themselves. Survival and liberty crucially depended on diverse cultural knowledge systems that would make tropical soils yield. Maroons could plant foodstaples presumably of their own choosing—even labor-intensive crops such as rice and cassava (manioc). All that stood between maroons and the exercise of African dietary preferences in tropical America was the availability of African seeds and rootstock.

THE QUILOMBOS OF MINAS GERAIS

The discovery of gold in the Minas Gerais region of southeastern Brazil in 1695 added a new dimension to the colony's dependence on slave labor. Demand for slaves was traditionally driven by coastal sugar plantations; now it expanded to the newly discovered goldfields of the mountainous interior. Fortune seekers poured in and mining operations proliferated along riverbeds and hillsides. Nuggets were captured by sieves, while gold dust was trapped by immersing livestock hides in running water (figure 5.1). The mines were immensely productive and profitable: over the course of the eighteenth century, Brazil created 80

FIGURE 5.1.   Illustration of gold washing, Brazil, ca. 1821–25.

SOURCE: Rugendas, *Viagem pitoresca através do Brasil,* pl. 3/22, following p. 166.

percent of the world's supply of gold. The discovery of diamonds in the 1720s added new impetus to the importation of African slaves, and the slave markets of west-central Africa continued to supply the demand.[4] Mining, like sugar, was carried out on the backs of enslaved laborers (figure 5.2, plate 4).

Food supplies in this new El Dorado were chronically short during the first frantic years of the gold rush. In the stampede to get rich, agriculture was largely ignored. Mine owners in the Minas Gerais region often denied slaves the right to plant provision grounds and frequently ignored the law allowing them Sundays and holy days for cultivating their own food.[5] Famine threatened on sev-

FIGURE 5.2.   Illustration of panning for gold, Minas Gerais, Brazil, by Carl
Friedrich Philipp von Martius, ca. 1817–20.

SOURCE: Spix and Martius, *Travels in Brazil,* opposite the title page.

eral occasions; predictably, slaves were the first to feel its effects. The brutal
labor regimes of the diamond and gold mines, compounded by food scarcity,
resulted in high rates of slave mortality.[6] Slaves repeatedly turned to flight. As
in other New World slave populations where food shortages were endemic,
hunger stoked the resolve to escape. The rugged montane terrain of Minas
Gerais offered the hope of freedom and safe refuge.[7] Mutually intelligible
African languages at times abetted flight from slavery. Many of the arriving
Africans were speakers of the same subfamily of Bantu languages, which meant
that they could understand each other with little difficulty.[8]

As escapes proliferated, maroon communities—known in Portuguese as
quilombos (from the Kimbundu word *kilombo*)—sprang up in inaccessible
mountain areas. The fugitives sited their hideaways inside forests and near reliable water sources. They used surrounding rock formations as defensive structures and for surveillance of approaching danger. By the 1720s, runaways from
gold and diamond mines in Minas Gerais had founded three major concen-

trations of quilombo settlements. The largest (estimated to have numbered six thousand people, including some Amerindians and whites) was known as Campo Grande. Its leader was reputed to be an African prince whose military defeat in his homeland led to his enslavement. Campo Grande was actually a cluster of quilombos in which satellite agricultural enclaves formed the periphery of a central and more defensible core settlement. The quilombos combined African and Amerindian agricultural practices and crops, emphasized self-sufficiency in food production, and worked the land together in common fields. They relied upon trusted commercial agents and miners as market intermediaries to trade gold dust, precious stones, woven goods, and livestock products (from cattle and sheep they raided) for other essentials.[9]

By the mid-eighteenth century, the Campo Grande quilombos increasingly were seen as obstacles to the expanded occupation of the area by ranching and mining interests.[10] Two of the enclaves, Ambrósio and São Gonçalo, were repeatedly attacked by paramilitary armies; the former was decimated in 1746 and finally destroyed in 1759 along with São Gonçalo. The militias sent against them made a point of burning community food fields and subsistence reserves. The ruins of these quilombo settlements, still very much in evidence in the 1760s, were mapped by a new expedition led by Inácio Correia Pamplona that returned to find maroons not taken during previous campaigns.[11]

Figure 5.3 shows the strategic siting of the Ambrósio enclave as recorded by the Pamplona expedition. It was located next to a swamp (#4) in the Serra da Canastra. Protected by the arms of a river tributary (#5) and guarded by a mountaintop sentinel (#2), the village compound was surrounded by a rampart and defensive trenches (#1, #6, #8), one of which was booby-trapped with staked pits. Access to the center was controlled by sentry posts (#3). Dwellings (#7) were located within the perimeter, agricultural land and pasture outside it (#9). The Pamplona expedition could not help but admire the tenacity of their maroon quarry, who had resettled the area after eluding capture a decade earlier. There, the military force saw a beautiful, extensive field planted to maize.[12]

The ruins of neighboring São Gonçalo (figure 5.4) provide insights into quilombo land use. Inside the defensive structures (#2, #4, #8) was a central residential area with dwellings (#6) and a food garden (#3). Located within the central ring were also granaries and structures vital to the community's survival: a structure with looms (#10) for weaving cotton and wool, another for ironworking (#1), and a hut with mortars and pestles (#7). An outer ring (#5), trenched for additional protection, likely held farmland and pasture. It was in turn surrounded by an unbroken arboreal perimeter, composed of natural forest (#9). This vegetative barrier kept livestock from straying and the entire community

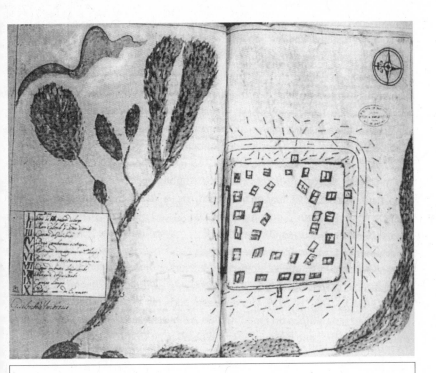

Legend:

1. Palisade and trenches
2. Mountaintop observation post
3. Sentry posts
4. Marsh with pits and stake traps
5. Gallery forest along streams
6. Defensive perimeter with traps
7. Quilombo houses
8. Defensive ditch surrounding quilombo
9. Agricultural land

FIGURE 5.3.   Quilombo do Ambrósio, Minas Gerais, Brazil, ca. 1769, drawn by a member of the Inácio Correia Pamplona paramilitary expedition after Ambrósio's initial destruction in 1746.

SOURCE: Biblioteca Nacional, *Anais da Biblioteca Nacional,* 111.

hidden from view. As with the other maroon encampments in Minas Gerais, the São Gonçalo quilombo combined agriculture with animal husbandry.

In nearby quilombos, the Pamplona campaign found sheep and cattle; fields planted to cotton, cassava, maize, and beans; looms, mortars and pestles, and mill presses for sugar extraction. The Campo Grande quilombos included many inhabitants who had been born there and had never known slavery. The militia was charged with enslaving their prisoners and starving those who man-

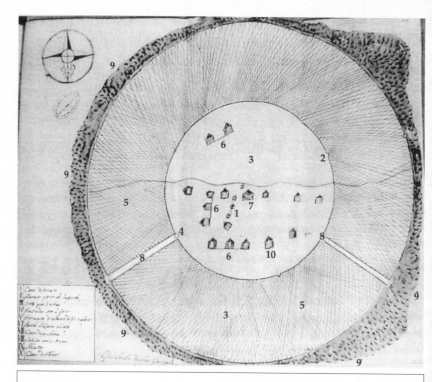

Legend:

1. Blacksmith structures
2. Holes for escape
3. Garden area
4. Entry with two booby traps
5. Trenches
6. Walls connecting houses

7. Milling structure with mortars and pestles for pounding grain
8. Exit with stake traps
9. Primary forest
10. Structure with weaving looms

FIGURE 5.4.   Quilombo do São Gonçalo, Minas Gerais, Brazil, ca. 1769, drawn by a member of the Inácio Correia Pamplona paramilitary expedition ten years after São Gonçalo's initial destruction.

SOURCE: Biblioteca Nacional, *Anais da Biblioteca Nacional,* 107.

aged to elude capture. They slaughtered the cattle and sheep, razed all structures, emptied granaries to supplement their own provisions, torched the grazing areas, and destroyed crops growing in the agricultural fields, lest they serve as food for any who escaped. These military reports indirectly suggest land-use practices and agricultural technologies of an African provenance.[13]

In the mid-eighteenth century, the majority of escapees living in the Minas Gerais quilombos would have been born in Africa. African technical knowledge undoubtedly helped them master the challenges posed by daily life. Although reliant for the most part on New World food crops, the remarkable application of African agropastoral farming techniques was crucial for maroon survival. Sheep and cattle provide through their manure the means to grow food on the same field year in and out, without loss of soil fertility. Land can be kept in continuous food production by rotating it seasonally between an agricultural field and pasture. Animal husbandry thus allowed maroons to maintain their hidden communities and so avoid the risk of detection that would likely attend relocation to new sources of arable land. Livestock formed the integral component of a survival strategy that enabled maroons to continuously occupy their secluded strongholds.

The Pamplona expedition's discovery of extensive planted fields, granaries filled with surplus food, and animal herds testifies to the remarkable subsistence achievements of the Campo Grande quilombos. Many of these techniques were not learned from Amerindians, who knew nothing about cattle, sheep, or goats prior to the arrival of Europeans and Africans. Nor did Africans have to look to Europeans for instruction in animal husbandry, as Africa supported, then as now, some of the world's most practiced herding peoples. The quilombo land-use strategies of montane Minas Gerais—like the African mortar-and-pestle milling devices found in the abandoned compounds—involved skills and technologies that were part of the cultural knowledge systems introduced to the Americas by enslaved peoples.

Over time the quilombos of Minas Gerais were mostly suppressed, but not all were eradicated. Today, south of the extinct Campo Grande enclave, another cluster of quilombos survives in the mountainous terrain of Minas Gerais.[14] The settlements date to the eighteenth century, when small groups of slaves managed to escape from the diamond and gold mines of the region. The exceedingly remote location (which is still not connected by a modern road) undoubtedly safeguarded the residents' freedom until it was legally granted in 1888 with the abolition of slavery in Brazil. In these centuries-old hamlets, one can today see the influence of African food preferences having taken root, undisturbed by the depredations of anti-maroon militias.

Today, a village elder named João Ribeiro is their community leader.[15] João is known throughout the state for his efforts to secure the property rights of quilombo communities, who still do not have title to the land their ancestors settled. João tells the story of his great-great-grandfather who was born in Africa and survived the Atlantic crossing on a slave ship where many perished. He was landed in Brazil and sold to work in the mines of Minas Gerais. João says that his ancestor spoke "Kongo" (Kikongo), the language he and other slaves used to communicate and plot their escape. When they did flee, the runaways joined the community into which João was later born.[16] In 1988, exactly one hundred years after Brazil abolished slavery, João's community at long last became eligible under Article 68 of the new Brazilian constitution to petition the federal government for legal recognition of their communal land rights.[17] There are over 135 remote hamlets seeking this formal designation today in the state of Minas Gerais. They are not alone: over two thousand quilombos throughout the country have filed similar petitions. This number would seem modest given that more than 40 percent of all slaves brought to the Americas went to Brazil.[18]

João's quilombo hamlet retains a remarkable degree of agricultural self-sufficiency. Each household continues to grow much of what they need for subsistence. Surrounding each family's home is a garden plot, where favorite food crops and medicinals are grown. In these plots are many plants native to the New World, but also a considerable number introduced from Africa. João's garden, for instance, contains Amerindian cassava, sweet potatoes, maize, peanuts, squash, beans, and chilies. These are growing alongside African yams, (Bambara) groundnut, black-eyed peas, watermelon, sorrel, sesame, plantain *(banana de São Tomé)*, and the bottleneck gourd.[19] João also plants bitter melon (or balsam pear, *Momordica charantia*), an African species that repeatedly appears among the medicinals cultivated by diaspora communities of the Americas. Guinea fowl are raised in the community. They are valued for their meat and eggs and are used in the practice of the Africa-based candomblé religion.[20] That the ancient agricultural traditions of two continents would share so equitably the soil in João's garden seems just, an apt symbol of how the foodstaples of Africa were integrated with the crops of Amerindian America.

ETHNOBOTANICAL KNOWLEDGE
OF AFRICANS IN THE AMERICAS

The first generations of Africa-born slaves certainly carried experiences, knowledge, and crucial skills from their homelands. Many already were accustomed to growing and preparing Amerindian staples such as maize, manioc, and peanuts, which European ships had carried to Africa in the sixteenth century. Although the New World environments Africans encountered contained many flora and fauna unfamiliar to them, there were other species that would have been recognizable because they belong to plant genera found in both the Old and New World tropics. Nineteen genera from fifteen botanical families occur in both Africa and Latin America.[21] Plantation slaves and fugitive runaways alike depended on ethnobotanical knowledge for nourishment, healing, and their collective survival.

Africans in the Americas experimented with plants from their immediate surroundings and incorporated many into their diets, healing, and religious practices. Escaped slaves acquired additional knowledge of New World species in their early and repeated interactions with Amerindians, for initial generations of enslaved Africans frequently worked and suffered alongside them. Whether as fellow slaves or runaways, through exchanges with native peoples, or through their own tropical knowledge systems, Africans adapted to New World environments. They grew Amerindian tropical foodstaples such as cassava and sweet potato, and they learned to identify wild foods and autochthonous medicinals of plant genera found only in the Americas. Africans in the New World also established plants and technologies inherited from Africa—such as rice and plantains, the mortar and pestle for milling grain, and familiar cooking practices.[22]

We see the cumulative significance of this fusion of African and Amerindian knowledge systems in the longstanding homeopathic medicinal tradition of the circum-Caribbean region. In these areas local people still rely on many plant-based cures ("green medicine") for the treatment of common ailments. In many areas of tropical America where Maroon ancestors won their freedom, their countrymen hold in high regard Maroon knowledge of the valuable properties of wild botanical species.[23]

Many slaves arrived in the Americas with medical knowledge and skills as healers. An early example is that of the slave Esteban, one of four survivors of the ill-fated Spanish Narváez expedition that shipwrecked off Tampa Bay, Florida, in 1527. A Moor of African descent, Esteban was born in Morocco, where he had been enslaved. He is known to us from the account of the epic journey written by one of the survivors, Cabeza de Vaca. It took the castaways nearly eight years to return overland across the southern flank of North America to Spanish settlement in northern Mexico. As they walked thousands of miles through hostile Indian territory, the linguistic and healing skills of Esteban enabled the survivors to reach New Spain (northern Mexico), located a continent away.[24]

Another documented example of the medical expertise of an African slave is from colonial New England. During a smallpox outbreak in 1721, the Massachusetts theologian Cotton Mather tested the treatment recommended by his

"Coromantee" slave, Onesimus. Onesimus had described to his master the African practice of taking fluid from a mild smallpox infection and introducing it to an incision made on the arm or hand. This technique, called variolation, would typically lead to a mild survivable form of the disease but would thereafter confer lifelong immunity. During the Massachusetts colony's smallpox epidemic, which claimed many victims, Mather followed Onesimus's instructions and inoculated his son, who grew ill but did not die. Even though Mather morally justified the institution of slavery because "Negroes . . . had sinned against God," he nonetheless could capitalize on the medicinal skills of an enslaved African.[25]

For some slaves, knowledge of a plant cure could lead to manumission. The most famous example is Quassi, an Africa-born man brought to Dutch Guiana as a slave in the early eighteenth century. Quassi was credited with discovering in 1730 the febrifuge properties of a tree found in the colony. The previously unclassified specimen came to the attention of Linnaeus in 1761. Of nearly eight thousand plants Linnaeus subsequently catalogued and named, *Quassia amara* is the only botanical species named after an enslaved person.[26] However, the honor bestowed upon Quassi should be historically situated within the broader ethnobotanical context of his life. Representatives of the same plant genus are also found in the region of tropical Africa (the Gold Coast), where Quassi was likely enslaved. The bark is still prepared into infusions for the treatment of fevers. Significantly, the tree thrives in old farm clearings, where agricultural land is left in fallow. A known healer, Quassi likely observed, and experimented with, a related species regenerating around the plantation landscape.[27]

Diasporic Africans used both Amerindian and African plants in many of the syncretic religions that developed in the Americas. Several New World species that are botanically related to known African medicinals are featured in the ceremonies of Afro-Brazilian candomblé. They are known by their African vernacular names.[28] This suggests that some enslaved Africans identified closely allied specimens of the same genus in the Americas and substituted the New World species for those known in Africa.[29] The African plants that did arrive in tropical America were also adopted. For instance, the deities of Africa-based religions in Cuba (known collectively as *orishas*) are often supplicated with specific plants of African origin. The special dishes prepared for the *orisha* include okra, black-eyed pea, and jute mallow *(Corchorus olitorius).* Sesame and guinea pepper *(Xylopia aethiopica)* are used in the *cazuela de Mayombe,* a clay vessel where the gods are confined.[30] Offerings of particular plants—many African—are also found in Brazilian candomblé ceremonies.

These offerings include sesame, tamarind, yams, black-eyed peas, African oil palm, watermelon, bitter melon *(Momordica charantia),* and the African guinea fowl.[31]

HISTORICAL CONSCIOUSNESS, MEMORY, AND LEGEND

Plantains, rice, yams, and millet are the cornerstones of traditional West African foodways, cultivated there for millennia prior to the arrival of Europeans along Africa's Atlantic coast. All were present at an early date as subsistence staples in plantation societies of tropical and subtropical America.[32] Found throughout the Western Hemisphere in diverse colonial contexts, these African staples were undoubtedly introduced at numerous times and places. Each crossed to the Americas on vessels departing western Africa and, in doing so, crossed geographical, political, and linguistic boundaries to augment the gardens and food fields of tropical peoples old and new, indigenous and forcibly migrated. Plantains, rice, yams, and millet figure prominently in the landscape of food fields first cultivated by maroons.

For most of these plants we can find historical records that credit Europeans with instigating their cultivation in the New World. For example, the Spanish friar who carried the banana plant to Santo Domingo from the Canary Islands around 1516; Dutch commercial agents who promoted millet as the ideal plant to feed thousands of Africans interned in the slave depots of Curaçao; and Portuguese mariners who allegedly brought rice from Asia.[33] For the tropical tubers, yam and plantain, we have no direct claims. Spanish chronicler Oviedo, in Hispaniola about 1514, summarized the dominant perspective on each of these African food crops when he wrote that yams were "chiefly grown and eaten by Negroes."[34] The presence of the new crop among recently arrived slaves to the island suggests the likelihood that enslaved Africans pioneered them.

Just as it had in the mountains of Brazil, marronage persisted throughout Amazonic South America. In the Guianas—an area encompassing British, French, and Dutch colonies—the humid rainforests and swamps that bordered the burgeoning sugar plantations of the coastal regions stood as a forbidding, largely inaccessible frontier wilderness to whites—and as a sheltering refuge to slaves wishing to escape. In these vast uncharted forests, plantation deserters were eventually able to achieve food self-sufficiency. The low-lying tropical environments of the Guianas, and the agricultural basis of the plantations that slaves fled, moreover proved favorable for establishing many African dietary staples. Thus, food systems of lowland tropical America offer another way of seeing the African influence on the collective subsistence and survival of maroons.

During the era of plantation slavery, Suriname (formerly Dutch Guiana; about the same territorial size as the state of Georgia) imported as many slaves as the entire U.S. South over a similar time span. Slavery in the colony was notorious for its oppressive demands on labor and the attenuated life spans of its bondsmen. No less cruel were the punishments meted out to those who attempted escape from the colony's sugar plantations (exemplified by the mutilated slave encountered by the eponymous hero of Voltaire's *Candide*). Nevertheless, many Africans enslaved in Suriname took the risk. Enough apparently succeeded, for by the 1670s official reports evince growing concern over the large numbers of runaways living in the rainforest interior. By 1684—less than twenty years after trading New Amsterdam (New York) for the nascent English sugar colony of Suriname—the Dutch colonial government had signed its first peace treaty with maroons.[35]

Maroon oral histories offer a view of crop introductions that contrasts decidedly with European claims. Rice figures significantly in the legends of these communities, while the stories themselves speak to its seminal status in Maroon culture. One example is the tale of Paánza, recorded by anthropologist Richard Price in Suriname. It has been told by generations of Saramaka Maroons who traditionally inhabit the upper watershed of the Suriname River. The tale of Paánza is instructive for understanding the significance of seeds, rootstocks, and plantation subsistence fields as food sources critical to the survival of the Saramakas' founding ancestors. Paánza is a female slave, who one day decides to flee as she is harvesting rice on a plantation. But before she runs, Paánza scoops up some grains of ripened rice and stuffs this seed rice in her hair. She escapes, and brings the grains to the maroons living in the forest. This, Saramaka descendants recount, is how they came to grow rice.[36]

Price has unearthed evidence that Paánza was not an imaginary heroine of folklore, but a real person. She was apparently the mulatto daughter of an Africa-born woman, who gave birth in Suriname about 1705. Price dates Paánza's escape to the period 1730–40.[37] Her mother had been forcibly migrated to Dutch Guiana in the late seventeenth or early eighteenth century, the period when the colony vastly expanded its sugar economy and accelerated the importation of enslaved Africans.[38] The extension of the plantation sector into the rainforest interior of the Guianas from the mid-seventeenth century gave slaves like Paánza greater opportunities to escape.

Price draws upon archival sources to tie the Paánza story to a series of slave revolts on Suriname River plantations that occurred during the last decades of the seventeenth century. One of these, Providence Plantation, was run by settlers of a utopian community of Labadist Protestants.[39] Notoriously cruel treatment on this sugar plantation provoked many slaves to escape, as related in this passage from a Saramaka oral history: "They whipped you.... Then they would give you a bit of plain rice in a calabash.... And the gods told them that this is no way for human beings to live. They would help them. Let each person go where he could. So they ran."[40]

Through skillful combination of archival records and oral histories, Price connects the uprisings and mass escapes to the founding of the Saramaka Maroons as a people. In this telling, the lack of rice is used to convey the overwork and hunger that motivated Saramaka ancestors to run away and form their founding community. In the story of Paánza, the Saramakas commemorate a dietary preference that abetted their struggle for survival. Together, both stories underscore the centrality of rice to Saramaka life and culture.

The tale of Paánza also draws attention to the instrumental role of a woman of African heritage in the cereal's geographic dispersal. It shares salient features of the common foundation narrative held by Maroons across northeastern South America: how an enslaved woman introduced rice by using hair to hide the grains, delivering them—in one story—from a slave ship to a subsistence plot, or from a plantation field to a maroon enclave in another.[41]

To this day rice remains the indispensable daily staple of Maroons in Suriname and French Guiana. Written accounts suggest that Maroon rice culture has changed little over more than two hundred years of observations. Here, rice is a woman's crop. Females sow the seed in rainforest clearings and harvest the crop by cutting the grain-bearing stalks with a small knife. They bundle the sheaves, carry them back to the village, and mill the cereal by hand with a mortar and pestle.[42] Anthropologist Sally Price has written that the cultivation and processing of rice is "the most important material contribution that women make to Saramaka life."[43] In each task, female Maroons are carrying out a cropping system that in fact closely echoes methods long practiced by women across West Africa's indigenous rice region.[44] Maroon rice culture represents more than the geographical extension of an African cropping system, however. Botanical collections made in Maroon communities of French Guiana have found the African rice species present among the types cultivated there. Indeed, *glaberrima* rice was actively maintained in Maroon food plots.[45]

Rice is the foundational component of the dishes that Maroons in Suriname have long prepared to commemorate important life passages. The cycle of an individual's existence closes with funerary offerings of cooked rice. It is also consecrated in "memory dishes" that commemorate ancestors.[46] Eighteenth-century Dutch historian J.J. Hartsinck recorded the Maroon practice of serving two rice-based meals during the mourning period that followed a person's burial.[47] The offering of food for the dead is similarly an ancient African practice that continued well into the period of transatlantic slavery. Symbolically perforated ceramic vessels and food remains that predate the beginnings of the trade have been found in funerary rituals and burial sites along the middle Senegal River valley. Food offerings inflect a funeral mass as depicted in mid-eighteenth century Kongo.[48]

The meaning of rice to the Maroons reaches beyond considerations of taste and the geographical transfer of African dietary preferences to New World plantation economies. The cereal simultaneously serves to articulate a memory of enslavement and deracination—of loss, separation, and forced exile from Africa—as well as one of freedom. It is through women that this collective memory is activated in oral accounts and ceremonial practices. A female ancestor carries precious African seed grains off a slave ship; another removes the grains from a plantation provision field and brings it to the Maroons. Women plant the seeds for subsistence, which enables others to free themselves from plantation slavery and join them. Grains of rice consistently form the coda of a historical consciousness that links the New World descendants of Africans to the continent of their ancestors.

MAROON FOOD PREFERENCES IN THE GUIANAS

Rice and other African food crops continually surface in colonial documents that pertain to maroons in the Guianas. In 1748 French Guiana, a captured maroon youth named Louis testified during his interrogation that his community planted in their forest gardens "manioc, millet, rice, sweet potatoes, yams, sugar cane, bananas, and other crops."[49] Four of the seven crops he identified (millet, rice, yams, bananas) were introduced from Africa. In Suriname, Saramaka communities that had been pacified by peace treaty were visited in the 1770s by Moravian missionaries who sought to convert them to Christianity. The missionary records also reveal an impressive array of African food-staples planted in maroon food fields. These include rice, plantains, bananas, yams, the African groundnut, okra, pigeon pea, watermelon, and sesame.[50] An 1810 military expedition against maroons of Demerara—the British sugar colony that bordered Suriname to the west—"destroyed rice crops sufficient to feed 700 persons for a year, as well as large amounts of yams, tanias, plantains, and tobacco." Despite the expedition's evident success, the crops were established again within a year.[51]

FIGURE 5.5.　Militia units setting fire to a Maroon village, detail of the Alexandre de Lavaux map of Suriname *(Generale caart van de provintie Suriname),* 1737.
SOURCE: Bubberman et al., *Links with the Past,* pl. 7, p. 127.

Confrontations with maroons are vividly rendered in an illustrated map made by the surveyor Alexandre de Lavaux during a 1730–31 eradication campaign in the Suriname interior. His eyewitness drawings combine in a single tableau details of different assaults against maroon strongholds in the upper reaches of the Saramacca River. Figure 5.5 shows militia units setting fire to a village, which is surrounded by a defensive wooden palisade. Also depicted are the efforts of maroon fighters to cover the retreat of their fleeing compatriots. Their bows and arrows, however, are no match for the advancing soldiers and their muskets.

Figure 5.6, another vignette from the Lavaux map, depicts the razing of another maroon enclave. Volleys of gunshot are fired at the fleeing maroons, and dogs set loose to track them down. Lavaux makes racial differences explicit in his drawings, such as the militia's enslaved porters and soldiers who, like the maroons, are depicted with cross-hatching to distinguish them from the whites.

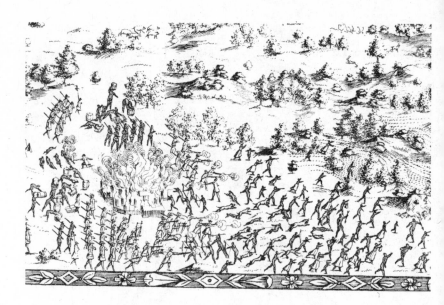

FIGURE 5.6.   Razing of Maroon enclave, detail of the Alexandre de Lavaux
map of Suriname *(Generale caart van de provintie Suriname)*, 1737.
SOURCE: Bubberman et al., *Links with the Past*, pl. 7, p. 127.

But one detail in the lower right corner of figure 5.6 catches the eye. Some maroon women are carrying pouches as they flee. The image raises the tantalizing
question of what items might have been prioritized in their desperate flight.
Could the pouches hold precious seed grains? This is certainly a reasonable spec-
ulation given that the ability of the maroons to survive the destruction of their
homes and food fields depended fundamentally on their capacity to plant food
in a reconstituted community.

Captain John Stedman was a Dutch-speaking mercenary officer who served
in several military actions against Suriname's maroons. During the years 1773–
77 his militia fought in the colony's northeast rainforest and swampy interior—
the area between the Cottica, Commewijne, and Marowijne rivers—where maroon settlements proliferated. Stedman wrote extensive eyewitness accounts of
his exploits, and his commentaries reveal the extent and productivity of the maroon agricultural fields he encountered.

Stedman noted that food was a priority to maroons fleeing the advancing
army; in particular, maroons made great efforts to carry off quantities of grain
and tubers, which, he observed, they grew in abundance. On one occasion, sol-

diers came across hampers of milled rice that were only abandoned when troops overtook the runaways carrying them.[52] Stedman's vivid descriptions of these skirmishes underscore how maroons prioritized food as they fled. But much was left behind in village granaries and unharvested fields. When not put to the torch, these abandoned reserves at times replenished the pursuers' own dwindling stocks. The militia commissary was so often depleted during these extended campaigns that soldiers were sometimes ordered to harvest the captured fields. It is in a maroon rice field that Stedman describes his first experience milling the grain by the traditional African method:

> He [the commander] gave out orders to subsist on half allowance, which he bid the poor men supply by picking rice and preparing it the best way they could for their subsistence. . . . It was no bad scene to see ten or twenty of us, beating the rice with heavy wooden pestles, like so many apothecaries, in a species of mortar, cut all along the trunk of a leveled purple-heart tree for that purpose, viz., by the Rebels, before they had expected to be honored by our visit. This exercise was nevertheless very painful, and verified the sentence pronounced on the descendants of Adam, that they should eat bread by the sweat of their brow, which trickled down my forehead, in particular, like a deluge.[53]

As in all maroon eradication campaigns, Stedman's militia was charged with capturing fugitives and denying any who escaped the means to resist or reassemble. In practice this meant burning maroon agricultural fields, homes, and defensive structures, destroying granaries and other food repositories—in effect, their entire subsistence base. Stedman notes that in just one campaign, soldiers demolished "more than two hundred fields of vegetable productions of every kind." In a diagram of one captured hamlet, he marked the community's substantial rice, cassava, maize, yam, and plantain food gardens (figure 5.7). The African crops he encountered in maroon food fields grew alongside Amerindian staples. Stedman even referred to the New World peanut by its African name when he noted its culinary importance to maroons: "The pistachio or *pinda* nuts they also convert into butter, by their oily substance, and frequently use them in their broths."[54] Other African crops that Stedman repeatedly identified in maroon food gardens included plantain, yams, pigeon peas, okra, and watermelon.[55] Stedman's catalog of crop destruction, however, is also a document of maroon achievement. It reveals unmistakably how Africa's botanical legacy had taken deep root in tropical America.

Maroon communities in the Guianas and elsewhere drew upon all the cul-

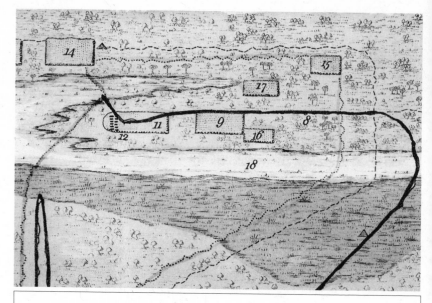

Legend:

8. Maroon defensive position
9. Rice and maize field
11, 16. Rice fields
12. Maroon hamlet

14. Old settlement
15, 17. Agricultural fields (cassava, yams, plantains)
18. Protective swamp

FIGURE 5.7.   Detail of a drawing of a Maroon settlement, northeast Suriname between the Cottica and Marowijne rivers, destroyed by the militia of Captain John Stedman in November 1776.

SOURCE: Stedman, *Narrative, of a Five Years' Expedition,* 2:128–29.

tural resources, knowledge, and skills belonging to their members. Liberty depended on the capacity to ensure daily subsistence. The synthesis of knowledge systems from the Old and New World tropics, initially forged in plantation food fields and then adapted to the swamps, jungles, and mountains of their secluded refuges, was indispensable for maroon efforts to remain free. The result was a new, robust plant assemblage that could be found in nearly all maroon communities of tropical America. It was a considerable subsistence base that included Amerindian cassava, maize, sweet potatoes, and squash, Asian sugarcane, yams (Old and New World types), *tannia* (New World *Xanthasoma* or Asian taro), groundnuts (peanuts and the African Bambara groundnut), beans (Amerindian and the African "peas"), peppers (chilies, as well as African melegueta and guinea pepper, *Xylopia aethiopica*), cultivated "spinach" (from the pantropical *Amaranthus* species and others), as well as bananas and plantains, rice, sesame, and okra introduced from Africa.[56] The African introductions in this tropical food complex form a significant component, and they reveal the reassertion of African dietary preferences where environmental conditions and freedom from bondage made it possible.

The Africanization of Plantation Food Systems

And he gave it for his opinion, that whoever could make two ears of corn or two blades of grass grow upon a spot of ground where only one grew before, would deserve better of mankind, and do more essential service to his country than the whole race of politicians put together.

JONATHAN SWIFT

1726

These plantation slaves received nothing from their master in the way of food or clothing, except only the small plot of land at the outermost extremity of his plantation land that he assigns to each slave.

JOHAN L. CARSTENS

St. Thomas, early eighteenth century

JEAN BARBOT WAS A FRENCH commercial agent who made two slave voyages from Africa to the New World. At the conclusion of his first voyage, to French Guiana in 1679, Barbot sold his cargo of enslaved Africans to sugar plantation owners and turned his attention to the plant and animal species that made tropical America a New World to him. He would later set down his observations in a book describing his experiences in the slave trade. Accompanying the text is an illustration of the botanical species that defined the surrounding landscape (figure 6.1). Barbot includes key Amerindian foodstaples: manioc, sweet potato, and the peach palm (valued for its fruit and palm hearts, known in his day as cabbages). He also shows two tropical fruits new to Europeans—the pineapple and papaya—and the dye plant annatto *(Bixa orellana)* that native peoples used for sunscreen, protection from insects, and as body paint and medicine. The European presence is evident in the depiction of the plantation crops, sugarcane and ginger (grown as a medicinal). But in an account that is otherwise silent on Africans except as troublesome trade commodities, Barbot illustrates one food plant that leaves us a botanical signature of their mounting pres-

FIGURE 6.1.   Illustration of plant and animal species found in Cayenne, late seventeenth century.

SOURCE: Barbot, "Description of the Coasts of North and South Guinea," in Churchill, *Collection of Voyages and Travels,* vol. 5, pl. 16, p. 200.

ence in the region. This is the plant he identifies as millet (lower left-hand corner), but is possibly sorghum, the larger of the African millets.[1] At the end of the seventeenth century, while maize was replacing sorghum in Lower Guinea, sorghum was being cultivated as a plantation foodstaple in the sugar economy of the New World tropics.

Barbot was not the first to record the presence of an African plant in the Americas. José de Anchieta (1534–97), a Jesuit missionary to Brazil, specifically noted African foodstaples that were being grown in Brazil and went so far as to commend them: "from Guiné there are many squashes and beans, that are better than those from Portugal."[2] During the sixteenth and early seventeenth centuries, other authors and artists describe or depict—within decades of plantation development—a cornucopia of African food plants they saw in tropical regions of the Americas: plantains, bananas, yams, the African groundnut, tamarind, pearl millet, guinea squash, pigeon pea, hibiscus, eddo, rice, sesame, okra, and the lablab bean.[3] These observations document, from an early period of European settlement, the many principal African food plants that were distinctly present in plantation societies of tropical America. This is perhaps unsurprising given their role as provisions on slave ships. But it is also a reminder that slaves not only grew plantation crops for their owners, they also cultivated plants for their own subsistence.

It is often forgotten that until the 1820s the majority of all immigrants to the Americas were born in Africa. Three out of every four persons who crossed the Atlantic before the nineteenth century came not from Europe, but in shackles from the continent of Africa. More than 40 percent arrived in Brazil.[4] Most were born in tropical Africa, and it was in the New World tropics where perhaps as many as 85 percent landed. Africans disembarked slave ships in tropical and subtropical environments that resembled in many ways those they had been forced to leave. They encountered landscapes with microenvironments suitable for growing traditional dietary preferences—yams, plantains, millet, sorghum, rice, and beans. They found vegetation from familiar plant families of pantropical distribution, whose potential food and medicinal properties awaited rediscovery. The botanical heritage of the African tropics, combined with that of Amerindians, laid the foundation for an experimental tradition that enabled slaves to confront a radically changed world that cared little whether they lived or died.

GUINEA'S PLANTS

African seed and root crops were pivotal to the collective efforts of enslaved Africans to secure an adequate subsistence in the humid and semiarid New World tropics. Many of these food plants were previously unknown to Europeans. Captains of slave ships first became acquainted with some as the victuals they loaded for their captives in the slave ports of the Guinea coast. African crops were no better known to plantation owners on the other side of the Atlantic, whose first experience with them was often in the food fields of their slaves. While many of the African foodstaples today are seldom appreciated as such, there is a linguistic bridge to their origins embedded in the place names given them by Europeans. The terms underscore their African provenance and the significance of the Atlantic slave trade for their diffusion.

European slavers distinguished many African crops new to them by using toponyms, or place names. The practice attached a geographical descriptor identifying where the plant came from (or could be found) to a general class of plant. For instance, many African plants are known by the toponym "guinea." Guinea is the common name Europeans gave to western Africa during the transatlantic slave trade, and it was applied in different languages and to a variety of novel foodstuffs loaded onto slave ships. To the British, for example, the common name for sorghum was guinea corn; the Portuguese used *milho da Guiné* interchangeably for pearl millet and sorghum. English also recognizes guinea grass *(Panicum maximum)*, guinea sorrel *(Hibiscus sabdariffa)*, guinea squash *(Solanum aethiopicum)*, guinea melon *(Cucumis melo)*, guinea pepper *(Xylopia aethiopica)*, and guinea bean *(Abrus praetorius)*—all present in the plantation period and today known to be of African origin.[5] Slave traders also adopted place names of specific regions and ports, where they purchased captives and provisions—Congo, São Tomé, Angola—for other African food plants. The pigeon pea, for example, was called Congo or Angola pea in English, *pois d'Angole* in French; the Bambara groundnut was known as Congo goober in English. The Portuguese also appended the place name *da costa* or *da terra* to a number of novel food plants they purchased along the African coast, such as *pimenta da costa* (guinea pepper), *palha da costa* (raphia palm), *obí da costa* (kola nut), and *banana da terra* (plantain).[6]

Species identified by such geographical toponyms underscore the importance of Africa-grown food to the transatlantic slave trade, but also the significance of subsistence to slavery's infrastructure. For the most part, the plants named in this way were African foodstaples. They include domesticates as well as Asian species introduced to the continent millennia earlier. The Portuguese first came across Asian taro, the banana, and plantain in western Africa, and they were duly signified with African place names. The African yam and Asian taro were indiscriminately referred to as *inhame da costa*—yam of the (African) coast—after the region where the Portuguese first encountered these important tubers.

The Portuguese borrowed the word "banana" from Gulf of Guinea languages; the Spanish, however, did not.[7] In Peninsular Spain both plantains and bananas were called *plátanos*. But in Spanish America the sweet fruit banana was linguistically differentiated from the plantain *(plátano)* as *plátano guineo*. The fruit is still known by its abbreviated common name, *guineo*, in former plantation areas of eastern Cuba, Colombia's Caribbean coast, the Pacific lowlands of Ecuador, and in El Salvador.[8] While European colonists would have little to do with the starchy plantain, they found the sweet taste of the banana pleasing.[9] The novel fruit is frequently featured in European paintings of plantation societies, such as in the still-life paintings that Albert Eckhout made of Dutch Brazil in the 1640s. Basque painter Joaquín Basarás, who lived in Mexico during the 1760s, added for the benefit of his audience the instructive label "Platano Guíneo" to his depiction of the fruit in one of his images of New Spain (plate 5).The geographical referent recalls the source of the plant's introduction to Spanish America.[10] The word *banana* was over time adopted by Spanish-speaking America for the sweet fruit type with the rise of North American fruit companies in the late nineteenth century.[11]

As the geographical descriptors underscore, the transatlantic dispersal of the African components of the Columbian Exchange did not occur in some unspecified past. The plants acquired their toponyms within the specific historical context of the Atlantic slave trade, applied with a deliberateness and calculation that betokens their significance to the trafficker. In this respect, African crop transfers were like none other in human history. In Atlantic Africa, the continent's foodstaples provided the captains of slave ships with provisions for the captives they carried to the New World. But not every slave voyage to the Americas exhausted its food reserves, as the Maroon rice narrative recounted in chapter 5 suggests. Slave ships arriving with leftover grains and rootstock gave the enslaved multiple opportunities to instigate cultivation of traditional dietary staples.

SUBSISTENCE AND EUROPEAN SETTLEMENT IN PLANTATION ECONOMIES

From the very beginning of European colonization of the Americas, subsistence and survival did not come without struggle. When initial food supplies ran out, settlers turned to the crops and knowledge held by other peoples. For example, the most important commemorative meal of the United States, Thanksgiving, famously celebrates the first communal feast of the Pilgrims and the Indians, who supplied the food that kept the Plymouth colony from starvation its first year. Indeed, few of the colonists knew anything about farming, and if the Pilgrims had landed early enough in the year to plant crops, they would have just as likely fallen on the generosity of their Indian hosts to survive. The Massachusetts Bay colony drew few settlers with a farming background, even if subsistence agriculture eventually prevailed. Food supplies were critical for immigrant survival. The Thanksgiving holiday symbolizes many things that are important to the American credo—fortitude, resilience, courage—but it also exemplifies the broad problem of subsistence that accompanied European colonization of the Americas.

Contemporary with the Massachusetts Bay colony, Providence Island off the coast of Nicaragua was also settled by English Puritans. The Puritans were again spectacularly unprepared for the subsistence challenges of this uninhabited tropical island, as historian Karen Ordahl Kupperman remarks: "They had to begin immediately producing crops for sustenance and export. Even experienced planters would have found this program difficult. For English gentlemen and servants, many unfamiliar with agriculture in any form, much less cultivation in the tropics, it was impossible."[12] Very early on, the colony organized agricultural production around African slave labor—not just to plant a cash crop, but also to produce food that would grow in the tropical climate.[13] By the 1630s slaves comprised nearly half the population. Eventually, efforts to establish an export crop were overtaken by a new plan to make the island a base for English privateering. Slaves were put to work growing the provisions and tending the livestock that would revictual the ships licensed to attack Spanish vessels. Among the crops were some that were African (plantains, muskmelon, "peas") in addition to cassava, beans, and "corn."[14] As in the Plymouth colony Puritan settlement, Providence Island depended vitally on the farming expertise of others, in this instance Africans, for food availability.

Settlement of the New World tropics posed a whole new array of challenges for European colonization. European immigrants were often town dwellers or tradespeople who held little experience in agriculture, much less food farming in the tropics. Few of their customary dietary staples grew in the tropical heat and humidity. Tropical climates do not support the temperate-zone crops that defined European foodways—wheat, barley, rye, and oats. To produce food in these alien environments, an entirely new assemblage of crops was required, including a different set of agricultural skills. But as a fundamental marker of cultural identity, dietary customs are stubbornly resistant to change, especially when immigrants are given a choice. European settlement of the New World tropics required either the wholesale adoption of unfamiliar foods or continuing dependence on the importation of customary foodstaples, an option unavailable to most colonists. For Africans, displacement to tropical America provided the environmental conditions receptive to the reestablishment of longstanding dietary preferences. The same environment also supported Amerindian staples that could make important contributions to slave subsistence security.

The use of enslaved Africans as plantation laborers in the New World overwhelms the important consideration that many were expert tropical farmers or herders. Among immigrants to the Americas, only Africans came equipped with knowledge of raising food in the humid and semiarid tropics. The native peoples of the Americas did not possess cattle, sheep, or goats until these Old World animals were introduced. The food systems that emerged on arid Caribbean islands, sometimes in conjunction with livestock keeping, also suggest African contributions. As enslaved Africans steadily replaced enslaved native peoples in the New World tropics, they brought their heritage of plants and agricultural practices to bear in the struggle against hunger and food shortages.

Africans filled the void left by native peoples who succumbed to or retreated from enslavement, disease, and genocide.[15]

Shortly after the English created a settlement on Barbados (1627), the colonists petitioned the mainland Dutch Guiana settlement at Essequibo for crops that could be grown for commerce and subsistence. Its Dutch governor encouraged a small band of Arawak Indians to accompany the Barbadians on their return voyage "to learn the English to plant." The Arawaks introduced Amerindian foodstaples to Barbados, but they were enslaved soon after their arrival. The instruction in tropical agriculture evidently had little benefit: four decades later, King Charles II in 1666 wrote that "Barbados and ye rest of ye Caribee Islands . . . have not food to fill their bellies."[16] This historical footnote reminds us that in the early plantation period white colonizers of tropical America knew little about growing food in the tropics, much less concerned themselves with its cultivation. But for the workforce they imported, subsistence was fundamental for surviving the rigors of plantation agriculture.

Enslaved Africans, many skilled in tropical farming, turned initially to the food plants Amerindians had domesticated, some of which, such as maize and manioc, were already part of African agricultural systems. But slaves also presided over the establishment of African dietary staples in plantation food fields. Writing of his residency in Barbados just twenty years after the island's transformation into a sugar colony, Richard Ligon explicitly credited slaves with the introduction of one important subsistence staple: "There is a Root, of which some of the Negroes brought the Seeds, and planted there, and they grew." He mentioned the plant's large size, that both the tuber and leaves were prepared as food, and the tuber's suitability as provision for lengthy sea voyages.[17] His description is likely of taro, a frequent provender of Middle Passage crossings. The enslaved Africans likely instigated the tuber's cultivation with rootstock they "brought" from the slave ship—tubers that remained from the slave voyage that carried them to Barbados. In 1648—just two decades into a plantation economy—Ligon observed other crops being cultivated for subsistence that we now know originated in Africa. These included plantains, bananas, the lablab bean, watermelon, tamarind, and purslane. Africans from most of the continent's ecological regions and farming systems were present in the colony in this early plantation period. Ligon noted that the slaves in Barbados were brought "from several parts of *Africa* . . . from *Guinny* and *Binny* [Bight of Benin], some from *Cutchew* [Cacheu], some from *Angola,* and some from the River of *Gambia.*"[18]

The problem of a dependable food supply also dogged English colonization of mainland North America. As in Barbados, indifference to food security would seem to be as much to blame as anything else. The corporate backers of the ill-fated Roanoke colony, founded in 1584 and conceived as a base for privateers to prey on Spanish treasure ships, "praised [its] supposed great fruitfulness, but recklessly compared it to the Garden of Eden."[19] Its first inhabitants were wholly unprepared for the subsistence challenges that awaited them. In the early years of the Jamestown colony (1607), settlers repeatedly failed to plant sufficient food: the "energies of the colonists had been largely absorbed in vain explorations for precious metals and other fantastic enterprises, as well as in the manufacture of pitch, tar, glass, and soap ashes." Trade with Indians and the intermittent arrival of ships with provisions barely kept them alive.[20]

The colony of Carolina, founded in 1670 by Barbadian planters (and their African slaves) and an early destination of English emigrants, similarly experienced severe food shortages in its first decade of settlement. Colonists relied upon food subsidies from England, and new settlers were asked to bring at least eight months' worth of supplies to better insulate them from chronic shortfalls. Sugar would not grow there, as it did in Barbados, and the colony skirted starvation while searching for a profit-making staple. Indian slaves, and the enslaved Africans who arrived with the early settlers, contributed critical farming and herding expertise. As Richard Ligon observed in 1648, slaves from Guinea-Bissau and Senegambia were represented among those carried to Barbados in its early settlement period, and among them would have been individuals who were skilled herders, rice growers, and tropical farmers.[21] Eventually, Carolina overcame its early subsistence struggles and even developed a profitable export economy based on foodstaples. In particular, Carolina supplied the British West Indies with three commodities—rice, cattle, and cowpeas—traditional mainstays of West African subsistence systems. After first establishing itself as a larder to the English Caribbean, the Carolina colony came into its own by the early eighteenth century as the Atlantic world's chief supplier of rice.[22]

However, it is in the Caribbean, and especially in the transformations imposed by sugar, where we see the extent to which Europeans depended on the knowledge of tropical peoples for food production. Before the arrival of Europeans, island Amerindians grew manioc (also known as cassava and yucca) along with sweet potato, maize, and other New World plants and combined them with fish and game to create diverse and nourishing diets. Manioc was described by Columbus (which he tellingly confused with the African yam); within a generation of his voyages, Amerindians were required to plant it on Hispaniola in vast provision grounds. Manioc, perfectly adapted to the climate and soils of the tropical Caribbean, its flour immensely durable in hot, humid conditions, soon became a major foodstaple for slaves, sailors, and soldiers.[23] The initial generations of colonizers used it as the basic ration for Amerindians enslaved in the placer mines and sugar estates that soon became the economic justification of empire in the region. Eventually, African staples—equally suited to cultivation in the New World tropics but with far fewer processing demands—began to replace manioc in the food fields of enslaved Africans throughout the Caribbean.

As island Amerindian populations were decimated by the "great dying" that followed contact and colonization, African slaves were imported to replace them.[24] By the early seventeenth century, the growing presence of new European powers in the region expanded the eradication of indigenous peoples to other Caribbean islands then coming under colonial control. Within a century of contact, Amerindians for the most part had vanished from plantation islands, aggravating a demand for slave labor that was increasingly filled by Africans. During the interval before their effective extinction in the Caribbean, indigenous peoples lived, worked, and died alongside enslaved Africans. Hunger and brutality were suffered in equal measures. Nevertheless, through these sustained interactions Amerindians imparted to Africans the knowledge systems and botanical acumen that facilitated survival in the New World tropics.

The labor regimes that Europeans forced upon Amerindians and Africans involved more than just the production of commodities for export. They also depended on the cultivation of food that would keep its unwilling participants alive. Both tasks fell to slaves. This included the common practice of making slaves entirely responsible for their own food availability.

Visitors to Caribbean sugar plantations during the formative development period repeatedly described the precarious subsistence situation of slaves, with food supplies inadequate and rations nonexistent. Writing about Hispaniola in 1561, when Africans numbered some twenty thousand persons, Juan de Echagoian observed that they were "not only corporally maltreated, but have so much work that they do not sleep at night and likewise do not eat, and in many plantations they are not given *casabe* [cassava] but beef and some bananas, and most of them go about in hides."[25] After many years of residence in the French Caribbean, Jean Baptiste du Tertre wrote in 1667 that slaves received "nothing save sufficient ground on which to plant food crops, build a house, and produce some other subsistence items."[26] Du Tertre also mentioned that slaves were given the appropriate time off for these endeavors, but the major intervention of the French Crown eighteen years later to redress problems of slave mortality belies his often salutary depictions of slave welfare. In the 1740s a St. Thomas planter similarly recorded how "slaves receive nothing from their master in the way of food or clothing, except only the small plot of land at the outermost extremity of his plantation land that he assigns to each slave." Correspondence from the Danish West Indies in 1718 added that "the negroes were expected to raise all their own food, except for such low-grade fish or defective Irish beef as might be allotted to them when the food supply ran short."[27] The burden of providing their own food added to the oppressive labor regime of sugarcane cultivation and effectively condemned slaves to perpetual hunger and untimely death.

A full day in sugar production left little time or energy for slaves to grow food and feed themselves, as George Warren noted after spending three years in Suriname during the 1660s. He left a telling portrait of planter indifference to the persistent malnourishment of their slaves:

> [They work] till Saturday afternoon, when they are allowed to dress their own Gardens or Plantations, having nothing but what they can produce from thence to live upon; unless perhaps once or twice a year, their Masters vouchsafe them, as a great favor, a little rotten Salt-fish. Or if a cow or horse die of itself, they get roast meat. . . . Their lodging is a hard Board, and their black Skins their Covering. These wretched miseries not seldom drive them to desperate attempts for the Recovery of their liberty, endeavoring to escape.[28]

Many sugar colonies had no laws specifying slaveholder obligations toward their bondsmen. Where they did exist, compliance was erratic or the standards pathetically low. The first generations of slaves in Brazil entered, in the words of historian Joseph Miller, "a notoriously mortiferous plantation environment" in their struggle "for sheer physical survival."[29] As early as 1604, the Portuguese Crown had attempted interventions on Brazilian sugar estates to ensure that slaves were fed. These had little effect, for the problem is repeatedly mentioned over the next century. As plantation owners increasingly left slaves in charge of their own subsistence, they faced opposition from Catholic clergy. Jesuit priests condemned the colonist practice of granting slaves one day in the week to grow and provide their own food—usually Sunday—because the time was both insufficient to produce adequate nourishment and interfered with attendance at mass.[30]

Most famously, the Code Noir of 1685 issued by the French monarch Louis XIV, formally decreed guidelines for treatment of slaves in his colonial empire. The edict compelled slave owners to ensure their bondsmen food allotments and to set minimal dietary standards. In Martinique, for example, each slave was allotted "two and a half pots of cassava per week."[31] Subsequent regulations increased the number of manioc tubers planted for slave subsistence and required that they be given plots to grow food. But the mandates were implemented in an expansionary period of sugar cultivation, when more land was given to cane at the expense of fields planted to subsistence crops.

While ostensibly motivated by humanitarian concerns, the Code Noir was poorly heeded. A report from Guadaloupe in 1704 observed that planters "feed their slaves badly, forcing them to work night and day . . . they furnish them with none of the necessities of life, and think about nothing but making sugar, which drives the Negroes to flee to the forest." A year later Father Jean-Baptiste Labat, a slaveholder in neighboring Martinique who styled himself a model plantation owner and an example to other planters, reported that island slaves were not given the time to prepare or eat their food during sugar season.[32] The intendant of the French isles, Charles-François Blondel de Jouvancourt, summarized in the early eighteenth century the common disdain of planters for the code:

> This slave mortality appears to be caused by the heavy labor that the planters make them perform without adequate nourishment. Some planters give them nothing except to let them work for themselves on Saturdays in order to earn their sustenance for the rest of the week; others give them only half the rations that are required by the ordinances of the king; and others give them even less. Still others give them neither the half rations nor the free Saturday. To be fair, there are some planters who give their slaves everything that is required by the Code Noir, but such planters are rare. The others, in contrast, are very numerous.[33]

Mandates to grow cassava (manioc) for slave subsistence ignored fundamental weaknesses of the crop as a food source in sugar economies. While manioc is well suited to tropical climates and thrives in depleted soils, its processing from tuber to flour or bread is time-consuming and laborious, requiring the leaching out of toxic alkaloids that naturally occur in the plant and are otherwise lethal (figure 6.2). For slaves made to plant manioc, the additional labor demands were onerous. Moreover, manioc packs little more than carbohydrates and alone cannot sustain human life. Slaves fed nothing but manioc would be condemned to slow but certain starvation. Thus, the edicts to ensure food availability set pitiably low nutritional standards and enforced a narrow subsistence regimen based on what could be grown cheaply and plentifully for food. Some planters gave their slaves

FIGURE 6.2. Illustration of manioc processing, French West Indies, by Jean-Baptiste Labat, 1722.

SOURCE: Labat, *Nouveau voyages aux isles de l'Amerique,* vol. 1, following p. 396. Courtesy of the John Carter Brown Library at Brown University, Providence, Rhode Island.

no food except for rations of rum, which presumably could be bartered for it.[34] While manioc initially emerged as the principal but monotonous foundation of that diet, in the evolving conditions of the sugar plantation economy, it gradually lost ground to more easily prepared foods from Africa.

AFRICANS AS CUSTODIANS OF AMERINDIAN KNOWLEDGE SYSTEMS

With the arrival of Africans in tropical America, the assemblage of subsistence crops was changing. African plants were quietly appearing alongside Amerindian staples in the food plots of plantation societies. Gonzalo Fernández de

Oviedo (1478–1557), who went to the West Indies in 1514, described guinea yams as a recent arrival to Hispaniola in the book he published in 1535. They were planted and consumed by African slaves.[35] Compelled to grow food for enslaved African gold miners in the early colonial period, Amerindians of the Colombian Chocó planted New World maize and the introduced plantain.[36] Writing in 1587 about Bahian sugar plantation development over the preceding twenty years, Gabriel Soares de Sousa mentioned several African staples among those that slaves planted for subsistence. These included guinea corn, yams, rice, and plantains.[37] In the seventeenth century, as the English, French, Dutch, and Danes founded plantation colonies in the New World tropics, a similar assemblage of African foods began to appear in slave food plots.

With the growing presence of enslaved Africans, the notable food and agricultural systems of Amerindians experienced an influx of new crops and farming techniques from tropical Africa. Enslaved Africans instigated the hybridization and intermingling of planting methods and foods that formed subsistence options in the New World tropics. On plantation islands where the native populations were annihilated, Africans became the custodians of the Amerindian botanical heritage.[38] This also included medicinal plants. The homeopathic healing tradition of the circum-Caribbean region, which retains its significance in medical practices to this day, is based on a plant pharmacopoeia that numbers more than six hundred species. For the most part of New World origin, this assemblage also includes important African medicinal plants established during the transatlantic slave trade.[39]

SUBSISTENCE TRANSFORMATIONS
WITH AFRICAN FOOD CROPS

The tubers introduced to the New World from Africa—yams, plantains, and taro—were ideally suited to slave subsistence plots because they demand little attention, are high yielding, and are readily prepared as food. Root crops hold a distinct advantage over cereals in hurricane-prone areas, as the underground edible part usually survives when struck by storms that routinely occur in the Caribbean. The tubers moreover can be continuously harvested, as needed. The fact that African yams *(Dioscorea cayenensis* and *D. rotundata)* send their roots deep into the soil structure also favored their establishment in the Americas, as this feature helped protect them from the reach of rooting pigs (an introduced faunal species), which easily unearthed the Amerindian tubers.[40]

In the early plantation period, the African yam and plantain revolutionized New World tropical food systems. Both were easily planted and required little

FIGURE 6.3. *Habitação de negros* (Dwelling of the blacks), by Johann Moritz Rugendas, 1820s.

SOURCE: Rugendas, *Viagem pitoresca através do Brasil,* pl. 4/5, following p. 205.

labor to cultivate and process into food; neither demanded the efforts needed to turn manioc into flour. The yam became so central to the regional food supply of the Caribbean that plantation provision grounds were frequently called "yam grounds." Yams produced prolific yields, stored well without spoilage, and could be cooked in many different ways. Their cultivation quickly spread throughout tropical America. But climate prevented the crop from thriving on the North American mainland, despite the efforts of the enslaved to establish the yam as a foodstaple. From the Carolina colony in the 1720s, Mark Catesby observed, "It is a Tropick Plant, not inclining to increase much in Carolina and will grow nowhere North of that Colony; yet the Negroes there . . . are very fond of them."[41] Nonetheless, the West African name *nyam* was used in the South to refer interchangeably to both the yam and the New World sweet potato. The African yam remains one of the most significant food crops that slaves pioneered in the Americas.[42]

The banana plant (both representatives of the *Musa* genus) is frequently featured as a shade and fruit tree in drawings of slave dwellings (figure 6.3). Both the plantain and banana were present at an early date in plantation societies.[43]

In colonies of New Spain, the fruit was known as *plátano guineo* or simply, *guineo*, which betokens its African source. The Portuguese and other Europeans gradually adopted the African vernacular term for the sweet fruit types. There is one documented introduction to the New World during this early period. Dominican friar Tomás de Berlanga carried cuttings of an unidentified clone from the Canary Islands to Santo Domingo in 1516. The European palate's attraction to the sweet banana would imply that Berlanga's clone was in fact the fruit type. Writing twenty years later, the Spanish court historian Oviedo averred that Berlanga's introduction was the *plátano*. However, the linguistic ambiguity of *plátano* makes it uncertain which cultivar—the sweet banana or cooking plantain—was meant.[44]

The plantain or cooking banana was such an indispensable foodstaple to Amerindian populations that many early European observers considered it native to the New World tropics. This assertion was soon dispelled. Alternative hypotheses, noting the plantain's true origins in Asia, attributed its arrival either to European newcomers or ancient trans-Pacific migrations by Polynesians.[45] The obvious corridor from Africa by way of the transatlantic slave trade has been ignored. In recent decades the plantain has been identified as an integral part of the Bantu expansion through equatorial Africa, and its importance as an ancient primary foodstaple of the region is no longer disputed. Plantains frequently provisioned slave ships leaving tropical Africa. The ships stocked the fruit and almost certainly the edible starchy stems. The perdurable corms easily survived ocean crossings and were soon reestablished by enslaved Africans who traditionally valued them as a dietary staple. Well into the nineteenth century, white elites in former plantation societies considered the plantain to be a staple of poor and "inferior" peoples.[46] One cultivar introduced from tropical western Africa to Brazil during the slave trade, the horn plantain or *pacova*, seemingly derives its name from *koba*, the African word for the plant in São Tomé and southeast Africa.[47]

Many attributes favored the plantain's importance as a foodstaple to enslaved Africans. It is a very high-yielding crop that can be harvested throughout the year. It is consumed raw, boiled, roasted, and stewed.[48] The plantain also provides a fermented drink, a process that Richard Ligon described in mid-seventeenth-century Barbados that was undoubtedly learned from enslaved Africans: "We pill off the skin, and mash them in water well boyl'd; and after we have them stay there a night, we strain it, and bottle it up, and in a week drink it; and it is very strong and pleasant drink, but it is to be drunk but sparingly, for it is much stronger than Sack, and is apt to mount up into the head."[49] Ligon illustrated both the plantain (figure 6.4) and banana plants, which had revolutionized

The Plantine

page 80

A Scale of :8:foote

Blossomd

FIGURE 6.4.   Illustration of the plantain tree, by Richard Ligon, ca. 1647–50.

SOURCE: Ligon, *True and Exact History*, opposite p. 80.

tropical Africa in previous millennia and to whose varietal development the ancestors of African slaves had made so many contributions.

For Ligon, an Englishman visiting Barbados during its critical transition to a plantation economy in the 1640s, the plantain and banana were new and exotic plants. But the island's enslaved Africans were well acquainted with the properties of each; in fact, the plantain was a preferred dietary staple. When the slaves "had Plantines enough to serve them, they were heard no more to complain," Ligon observed, "for 'tis a food they take great delight in." The "Negroes chuse to have it green, for they eat it boyl'd, and it is the only food they live upon."[50] He considered it "a lovely sight to see a hundred handsom *Negroes,* men and women, with every one a grasse-green bunch of these fruits on their heads, every bunch twice as big as their heads, all coming in a train one after another.... Having brought this fruit home to their own houses, and pilling off the skin of so much as they will use, they boyl it in water, making it into balls, and so they eat it."[51]

Observing that slaves did not similarly esteem the banana as a dietary staple, he attributed their bias to an aversion to sweetness. "This fruit is of a sweeter taste than the Plantine; and for that reason, the *Negroes* will not meddle with them, nor with any fruit that has a sweet taste."[52] The problem was not, as Ligon casually maintained, inherent African disdain for sweetness. Rather, an important distinction was being made: the cooking banana provided the starchy foundation for a meal, the sweet uncooked fruit did not, and any attempt by slaveholders to substitute one for the other in their diet would not be accepted.

The *Musa* plant's establishment in the New World tropics did not occur through the form that we in the modern world now encounter bananas and plantains, as a sweet fruit or starchy vegetable staple encased in its peel. Instead, its introduction to the Americas would presume a process similar to that which occurred during the Monsoon Exchange. Just as ancient mariners carried the plant's edible vegetative stems across the Indian Ocean to Africa, so did European vessels likely transport the rootstocks of bananas and plantains to the Americas.

Plantains were frequently stowed as food on slave voyages. But transported by the familiar bunch, they could not have served as planting stock. The historical record does not indicate whether the *Musa* plant's edible root stems were fed to slave-ship captives. But given their historical use in equatorial Africa as a famine food, and exceptional viability when stored for long periods of time, their possible role as victuals cannot be discounted.[53] The pseudostem has been reported as provender for cattle and pigs in Africa and the Canary Islands.[54] Perhaps pseudostems, suckers, and corms were similarly used to feed the live animals that slave ships routinely carried as fresh meat for their crews. Ligon suggests this possibility in his description of how slaves used *Musa* cuttings to feed swine: "When it [the fruit] is gathered, we cut down the Plant, and give it to the Hoggs, for it will never bear more. The body of this plant is soft, skin within skin, like an Onyon, and between the skins, water issues forth as you cut it. In three months, another sprout will come to bear, and so another, and another, for ever; for we never plant twice."[55]

Ligon's commentary on the plantain and its multiple uses, often expressed through the collective "we," suggests the inventiveness of the island's colonists, but in fact describes practices familiar to an enslaved workforce intimately acquainted with the plant's properties. Slaves remain silently in the background of Ligon's celebratory account of plantation development on Barbados. Typical of the chroniclers of his era, he sweeps the uses of novel plants and newly learned practices under the enterprising activities of Europeans, who were in the process of remaking the New World to suit their commercial purposes.

The plantain continued to be considered "Negro food" in the circum-Caribbean region into the twentieth century.[56] The vernacular names for the *Musa* species in the New World tropics provide one indication of its African heritage in the Americas, as do the ways in which African slaves used the plant's leaves, fruits, and underground stems. In this way we begin to see the significance of Africans in providing the foundational knowledge for the establishment and widespread dissemination of *Musa* plants in tropical America.

THE ANTILLEAN SUBSISTENCE SYSTEM

Writing of the Danish Caribbean in 1768, Moravian missionary C. G. A. Oldendorp observed that "the kind of agriculture practiced in these islands is quite different from the European variety. With the exception of several areas on St. Croix, the soil is neither turned up by the plow nor dug up by the spade. The hoe alone is used for this purpose."[57] This striking feature of African agriculture was identified and drawn by the Capuchin missionary João António Cavazzi de Montecúccolo, who went to Angola in 1654 (figure 6.5). The long-handled iron hoe, the fundamental implement of African agricultural systems, became important as well in plantation societies. Even though the labor of twenty slaves was needed to do the work of "one man and boy with two horses," the European preference for plow agriculture was not widely employed on plantations. The African method was instead predominant. Characteristically, many types of hoes were used: small-handled ones for ridging and mounding the soil, long-handled types for breaking up the ground and preparing it for planting. Hoe agriculture remained the principal way that slaves and their descendants in postslavery societies prepared their food plots (figures 6.6, 6.7).[58]

Cultivation with handheld hoes, rather than draft-animal traction, represents the transfer of African farming methods, as Oldendorp implicitly understood. It is among the agricultural practices whose distinctive character has been re-

FIGURE 6.5. Illustration of Angolan woman with hoe, by João António Cavazzi de Montecúccolo, ca. 1660.

SOURCE: Cavazzi de Montecúccolo, *Descrição histórica dos três reinos,* 39.

ferred to as the Antillean subsistence system, after the alternative historical name for the Caribbean islands.

The Antillean cultivation system shares many features with subsistence agriculture in mainland tropical America, but differs in key respects. For example, root crops have a distinct advantage over cereals, especially on Caribbean islands frequently lashed by hurricanes. The food source of a tuber is underground, and so is usually spared when high winds strike. As a result, New World root crops (manioc, sweet potatoes, arrowroot, yams) tend to receive more emphasis in Caribbean food production strategies; this was also true for Amerindian populations.

Another characteristic of the Antillean system derives from the greater intensity of the African population influx in the Caribbean and the early extinction of native peoples on the islands. The overlap between Africans and Amerindians in the Antillean system began and ended in just a few generations, in contrast to a longer and more sustained period of interaction between the two populations on mainland tropical America. In the Caribbean—where more

FIGURE 6.6. Adelaide Washington with hoe and basket on head, St. Helena Island, South Carolina, early 1900s.

SOURCE: Photo by Leigh Richmond Miner, in the Penn School Collection. Courtesy of Penn Center, Inc., St. Helena Island, South Carolina.

FIGURE 6.7.   Luisa Carvalho in former rice plantation region, Pará, Brazil, 2002.
SOURCE: Photo by Judith A. Carney.

than 40 percent of Africans disembarked—slaves reached a much higher proportion of total population and density per land area than was found almost anywhere else in the Americas. A comparative perspective reveals identifiable features of the Antillean food system, making it in part possible to disentangle specific historical contributions of enslaved Africans from the broader Amerindian influence on New World tropical farming systems.[59]

Researchers Riva Berleant-Schiller and Lydia Pulsipher identify a core plant assemblage that characterizes the Antillean subsistence system at its broadest level. Of the eight core crops that form the foundation food-crop complex, half are introductions from Africa (yams, the plantain/banana, taro, and pigeon peas), the remainder Amerindian.[60] Africa's pigeon pea assumes a primary role in subsistence and land-use strategies because of its importance as a plant-based protein source, the use of its shells for animal feed, and the role of the legume in restoring soil fertility. The pigeon pea is also used as a plot boundary marker. In many parts of the Caribbean it is considered a medicinal and is ceremonially used in Afro-syncretic religious practices.[61]

These contributions to the New World tropics are obscured on the mainland, where Amerindian populations persisted well beyond contact. But in the Caribbean, African slaves built their survival in part on their African heritage and on what the first generations of their enslaved forebears learned from vanishing Amerindian populations. The rapid collapse of Caribbean Indian populations on many sugar islands left Africans and their descendants as the living repository for the accomplishments of two significant tropical farming systems.

The African contributions have been muted because of the entangled histories of both peoples in the Americas and an unreconstructed view that portrays slaves as little more than muscles of European ingenuity. African plants accompanied one of the most significant population movements in human history—an involuntary migration that forced the departure of millions from the African continent. In emphasizing the geographical areas where food crops important in the Atlantic slave trade were domesticated, Columbian Exchange scholarship unintentionally occludes the African components of intercontinental crop exchanges and the role of Africans in pioneering them elsewhere. The fact that the plantain, banana, and taro originated in Asia does not diminish the significance of these crops, and African experimentation with them, to African food systems that existed long before the arrival of Europeans. Instead, an invigorated approach engages the critical role of these crops in African foodways, as provisions carried on slave ships to the New World, and the meaning of these and other African dietary staples to a continent's deracinated peoples. It draws attention to the desperate circumstances that enslaved Africans faced in feeding themselves and the ways that these plants abetted their survival and eased their despair.

In the context of Atlantic slavery, food encapsulates two opposing perceptions. To the slave trader and owner, it was a cost overhead, a necessary input to a specific, commercial end. To an enslaved person, it represented sustenance and the means of survival. African foodstaples were transported across the Middle Passage as expedient provender to keep captives alive until landed. But to Africans in the New World, these same staples embodied both potential nourishment and dietary preferences vested with identity and cultural meaning. The enslaved person viewed these foodstaples as components of familiar foodways; the slaveholder treated them as a cost of business.

African crop transfers thus were unlike any other in the history of intercontinental botanical exchanges. They took root in the subsistence fields of plantation economies wherever the conditions to grow and process them permitted. African foodstaples were pioneered by enslaved, underfed Africans who faced in the early colonial period the real prospect of starvation. In their struggle to stay alive, enslaved Africans drew deeply upon the agricultural expertise and the crops of their own heritage, while adopting the knowledge systems and plants bequeathed to them by Amerindians. Through this new assembly of tropical food plants, we grasp in part the remarkable efforts they made to stay alive in the face of despair and unimaginable toil.

SEVEN

Botanical Gardens of the Dispossessed

In the same way that Europeans brought to America plants and seeds that they judged beneficial, so too did the Africans.

WILLEM PISO
on sesame, guinea squash, and okra in Dutch Brazil, 1648

Bunched guinea corn [sorghum]. . . . But little of this grain is propagated, and that chiefly by Negroes. . . . It was first introduced from Africa by the Negroes.

MARK CATESBY
Carolina, 1754

Cola africana—an African fruit, introduced by the Negroes before [Hans] Sloane's time [1687–89], called bichey or bessai.

BRYAN EDWARDS
Jamaica, 1793

THE HISTORICAL RECORD ON THE question of African plant introductions to the Americas is not so silent as we might suppose. A salient footnote of the plantation period is the number of European accounts that actually credit slaves with the introduction of specific foods to the Americas, all previously grown in Africa. These accounts were mostly written by planters and naturalists of different nationalities, working in colonies throughout the Caribbean and North and Latin America.

These European commentaries inadvertently attribute an agency to enslaved Africans that contrasts with colonist narratives of how slaves came only with their bodies, bereft of farming skills and knowledge, and were given these skills and knowledge by white planters. They also conflict with later claims of planter inventiveness in introducing new crops suitable for cultivation that were in fact longstanding African foodstaples. The commentaries identify Africans as pioneers of crops entirely novel to their masters. They lend support to the significance of slave food fields as staging grounds for the "diaspora" of African food plants. In these sites of subsistence, slaves instigated the cultivation of many dietary preferences without planter guidance or coercion.

Through such accounts, we can identify at least a dozen plants whose introductions to the New World are directly attributed to African slaves. For example, Willem Piso, a naturalist who worked in Dutch Brazil in the 1640s, made drawings of *belingela,* the African eggplant known at that time in English as guinea squash; he asserted that it was introduced by Angolan slaves, along with okra and sesame.[1] Of the bonavist or lablab bean, Piso's scientific collaborator Georg Marcgraf wrote, "This plant was brought from Africa to Brasil."[2] Sir Hans Sloane, founder of the British Museum who was in Jamaica from 1687 to 1689, wrote of a bean "brought from *Africa*" that he described as "almost round white Pease something resembling a kidney with a black Eye not so big as the smallest Field pea."[3] This "calavance" pea, so clearly strange and new to him, is the first certain description of the African cowpea in English America, which became known in the colonies as the black-eyed pea, after its distinctive appearance.[4] In the Carolina colony, English naturalist Mark Catesby attributed to slaves the introduction of sorghum and millet.[5] French botanist François Richard de Tussac (1751–1837), who worked in the colonies of Saint Domingue and Martinique, credited slaves with bringing the cytisus (pigeon) pea to the French Antilles. British historian John Oldmixon (1673–1742), echoing Oviedo's account of yam introductions to Hispaniola two hundred years earlier, contended that yams "were brought thither [to Barbados] by the Negroes."[6] Luigi Castiglioni, an Italian botanist, wrote during his travels in the United States (1785–87) of a plant that "was brought by the negroes from the coasts of Africa and is called okra by them."[7] Naturalists Johann Baptist von Spix and Carl Friedrich Philipp von Martius encountered okra in early nineteenth-century Brazil and wrote, "It seems to have been introduced by the negroes from Africa."[8] Thomas Jefferson claimed that sesame "was brought to S. Carolina from Africa by the negroes."[9]

How is it possible that slaves could have introduced all these plants when they were brought to the Americas with few or no personal belongings? Could so many historical commentaries be wrong?

The introduction of African crops is coincident with the arrival of enslaved Africans who had subsisted on these foods during the Middle Passage.[10] Hans Sloane touched on the crucial connection between African plants and slave ships when he reported that the African kola nut arrived in Jamaica from "seed brought in a Guinea ship."[11] Similarly, the peanut—a plant actually of South American origin that was introduced to western Africa in the early sixteenth century—made the journey, Sloane writes, on slave voyages "from *Guinea* in the *Negroes* Ships, to feed the *Negroes* withal in their Voyage from *Guinea* to *Jamaica.*"[12] Captains of slave ships loaded these crops and other African food-staples onto their vessels to facilitate the transportation of human beings—the kola nut to make spoiled water palatable and the food crops to provision captives during the Atlantic crossing. Once landed, captains of slave ships dispersed their human cargoes and returned to the metropolitan centers of their financial backers. But during the time they remained in port, the occasional surpluses of African seeds and rootstock were likely removed from their vessels. In the early colonial period the principal food crops of the African continent had already become a botanical feature of New World plantation landscapes.

Plantation owners first encountered African crops in the food fields of their slaves. In these subsistence plots (also called "Negro plantations" in historical accounts), Europeans noticed a different assemblage of plants, some of which were entirely new to them. Among the ones Willem Piso identified in seventeenth-century Brazil were okra, sesame, and guinea squash. Hans Sloane added that Jamaican slaves also grew millet, sorghum, rice, and the black-eyed pea in their food plots. Millet "is to be met with in some *Negro's* Plantations," he wrote, "though not so commonly as the former [sorghum]." He found the black-eyed pea "planted in most Gardens, and Provision Plantations, where they last for many years."[13] It was in slave food plots that Mark Catesby observed the cultivation of sorghum and millet in early eighteenth-century Carolina: "These two grains are rarely seen but in Plantations of Negros, who brought it from Guinea, their native Country, and are therefore fond of having it."[14] Jefferson attributed the presence of sesame in the Southern colonies to slaves "who alone have hitherto cultivated it in the Carolinas & Georgia."[15]

These commentaries illuminate a shadow world of cultivation that had evolved in the struggle of the first generations of enslaved Africans to ensure food availability. This cultivation system was closely tied to subsistence and emphasized crops that Africans established by their own initiative. It stood in stark contrast to the commodity fields of the plantation, where slaves grew crops under the directives of the slaveholder. In the interstices of the plantation economy, slaves cultivated subsistence crops on many types of food plots. They planted dietary staples that were particularly suited to tropical and subtropical growing conditions and that were often ones they preferred.

It was in "Negro" food plantations and in the yards around slave dwellings where the African components of the Columbian Exchange made their initial New World appearance. European naturalists and slaveholders encountered these new food crops in slave subsistence sites, acknowledging in their comments the role of enslaved Africans. The significance of the African plant assemblage for slave survival, and the cultural funds of knowledge that informed their establishment as subsistence staples, was not lost on Willem Piso in seventeenth-century Brazil. Assessing the fundamental dietary role of a wide range of new tropical food plants that he found in slave food plots, Piso remarked, "In the same way that Europeans brought to America plants and seeds that they judged beneficial, so too did the Africans."[16]

PLANTATION FOOD FIELDS

In the early period of plantation development, food supplies were chronically scarce. At the end of a rigorous day in plantation labor, food availability depended very much on the extra effort slaves could muster to provide for their own subsistence. One Dutch traveler to Virginia reported in 1679 that the tobacco estates demanded long hours of toil "as if planting [tobacco] were everything." After an exhausting day, the slaves "still had to pound maize for their food, which was mostly hominy and poor in meat."[17] Acknowledging that the time to tend their food plots counted for just a minor portion of each day, Johann Martin Bolzius observed in Georgia's early plantation period, "If the Negroes are Skilful and industrious, they plant something for themselves after the day's work."[18]

Overwork and fatigue affected the ability of enslaved Africans to cultivate some preferred foodstaples. Sloane wrote in the 1680s, for instance, that hand-milling made rice more labor-intensive than other subsistence options. "This grain is sowed by some of the *Negro's* in their Gardens, and small Plantations in *Jamaica,* and thrives very well in those that are wet, but because of the difficulty there is in separating the Grain from the Husk, 'tis very much neglected, seeing the use of it may be supplied by other Grains, more easily cultivated and made fit for use with less Labour." Even so, the pottery that Jamaican slaves produced in the seventeenth century included decorative impressions made with rice grains.[19]

The labor of plantation slaves was vested not only in the production of export crops, but in the building of the estates themselves. Plantations were carved from forests and swamps cleared by slaves. As only hand tools were available to accomplish this task, the removal of even a single tree demanded considerable work. The stumps and underbrush had to be burned and uprooted before the soil could be cultivated. The physical demands on enslaved workers in the initial period of plantation development made the supply and quality of food vital to their survival. Even in French Catholic colonies where the Code Noir and other metropolitan edicts aimed to enforce planter compliance, slaveholders consistently failed to provide adequate food to their enslaved workforce. Whether for reasons of planter indifference or economic efficiency, food availability depended critically upon the efforts of those already consumed by plantation work to provide for their own dietary needs.

Enforced self-reliance in plantation economies nonetheless had an unforeseen consequence. Arrogation of subsistence obligations to slaves generated both the need and the opportunity to establish many African foodstaples in their provision grounds and dooryard gardens. In pioneering these subsistence foods, slaves strengthened their own food security, diversified the dietary options otherwise available to them, and reinstated some traditional food preferences. These crops gave enslaved Africans the opportunity to choose in part what they consumed, beyond what a plantation owner deemed appropriate or convenient to feed them. Through this narrow window entered the botanical legacy of a continent.

SITES OF SUBSISTENCE

Enslaved Africans established their subsistence staples in three distinct settings: in the individual plots that some plantation colonies granted them, on provision grounds, and in the yards surrounding their dwellings. The relative emphasis on each subsistence practice varied between plantation regions and over time. In the initial period of Brazilian sugarcane development, some slaves were already exercising the right to grow food on individual plots. Gabriel Soares de Sousa referred to these plots in the 1570s–1580s when he mentioned food-staples that slaves planted.[20] Writing about Dutch Brazil in 1646–48, Pierre Moreau described them as "little pieces of land on which, during the limited time they have for rest (after a twelve-hour day) they sow peas, beans, millet, and maize."[21] The subsistence convention of independent production eventually diffused from Northeast Brazil to parts of the Guianas and some Caribbean islands, where it became known as the Brazilian or Pernambuco system, named for the plantation society where it was initially implemented.[22]

As European slave traders had observed in Guinea over the same era, independent production was initially conferred as a privilege, not as a right universally granted to the slave population.[23] In Brazil, Jesuit priest André João Antonil reported in 1711 that it was given to senior slaves, usually those who held their master's trust. But in the context of chronic food shortfalls, placing slaves in charge of their own subsistence freed planters from the necessity to provide them food rations. Eventually the practice was expanded to include most of the enslaved male population.[24] On the Caribbean island of Barbuda, a 1715 inventory instructs that specific slaves be permitted to tend "their provision grounds without interference." The crops they grew included guinea corn, yams, maize, and legumes (possibly pigeon peas).[25] Even so, access to subsistence plots did not ensure slaves the requisite time to produce their own subsistence. They were usually allowed no more than a day a week to work their individual food plots, sometimes just a portion of one weekend day.

Brazilian slaveholders regarded the convention of independent production as a bestowed privilege, easily conferred but just as easily revoked. Slaves considered it a right, as their subsistence security depended upon it. Once it was granted, they resisted any attempts to withdraw it, as two documents first published by historian Stuart Schwartz illuminate. On a Bahian sugar plantation in 1790, a group of runaway slaves put forth conditions for their surrender. The demands suggest that the fugitives had been accustomed to growing food on their own subsistence plots and had been allowed to bring new land into cultivation. They listed as a condition for returning to plantation labor "the days of Friday and Saturday to work for ourselves not subtracting any of these because they are Saint's days." They also demanded "the right to plant our rice wherever we wish, and in any marsh, without asking permission for this."[26]

Apologists for the Pernambuco system liked to portray independent production as an example of planter benevolence and the inherently civilizing virtues of slavery. These plots after all gave slaves some degree of economic autonomy by way of new opportunities to produce marketable surpluses. In 1784 Brazilian naturalist Alexandre Rodrigues Ferreira opined that independent production provided slaves the means of legal manumission:

> Some owners of sugar mills have the custom to grant each slave as much land as he needs, according to his being single or married, and one or two days each week, in order that he can cultivate his plot. From the latter they obtain the yucca [manioc] flour, the maize and the beans for their sustenance and that of their wives and children, thanks to these days they work for themselves each week. . . . And the fact is, as we know by experience, that not only do they cultivate the yucca flour for feeding themselves, but also they are able to sell almost all agricultural goods as well as many domestic animals, until they accumulate enough money to free themselves and their children.[27]

Ferreira's idyllic characterization of independent production in eighteenth-century Brazil is at odds with repeated observations of the chronic inanition of the country's slaves over the more than three centuries that the institution endured there. Sugarcane production left slaves little time to work their subsistence plots. Nor did economic autonomy through the sale of marketable surpluses generate the revenue by which large numbers of slaves freed themselves. In fact, harsh labor and poor nutrition contributed to the persistent failure of the Brazilian slave population to increase naturally—even after the transatlantic traffic ended there in 1851.[28]

Historian Barry Higman enables us, however, to place Ferreira's comments in broader perspective. Writing about the attendant ill health, morbidity, and premature death of sugar workers in the British Caribbean, he has estimated that field slaves worked in sugarcane on average some thirty-five hundred hours annually, or about sixty-seven hours per week. This is equivalent to a modern worker accustomed to a forty-hour work week compressing (without time off) an *additional* thirty-five and a half weeks into a fifty-two week year. These figures do not include the extra time that slaves may have spent on subsistence cultivation. The result was an exhausted labor force whose life expectancies remained low.[29]

Under a sugar labor regime, only the healthy and physically fit could realize the surplus potential of independent production. But the convention of granting slaves an individual plot was a convenient one for planters, as it relieved them of responsibility for the dietary needs of their workforce. Moravian missionary C. G. A. Oldendorp made this point when he wrote about the practice on Danish St. Croix in the mid-eighteenth century: "From this [plot], they are to produce their own means of sustenance. The yield is generally great enough that it provides the diligent cultivator with a surplus beyond his basic needs, and from this he can provide himself with other commodities. This arrangement relieves the master of any further cares concerning the slaves than when the essentials for their sustenance are handed to them in kind, as is the case on several English plantations on St. Croix."[30]

Independent production was not the only subsistence system practiced in the early plantation period. In other areas, a portion of plantation land was set aside for the purpose of planting food crops for all the resident slaves. The cultivation of African yams for sustenance on food fields was in fact so common in many parts of the Caribbean that "yam grounds" became a metonym for provision grounds. Many of the southern colonies of mainland North America also adopted the practice of designating specific grounds where slaves grew their collective sustenance.[31]

However, by the eighteenth century the emphasis on provision grounds for food production was shifting on some sugar islands. The demand for sugar in European metropoles had grown to such an extent that even more land was brought into production. On islands of flat topography, cane cultivation eventually stretched from shore to shore. The acreage in food cultivation severely declined. Alexander Hamilton, born on Nevis but who lived in St. Croix until the mid-1750s, noted the erosion of subsistence security when he recalled that planters "appropriate only small portions to the purpose of raising food. They are very populous, and therefore the food raised among themselves goes but little way."[32]

Mountainous sugar islands such as Jamaica responded to the insatiable demand for sugar land by relocating food production to hilly areas, on soils that were marginal for cane, and to other less accessible locations. Although plantation owners sometimes subdivided the distant provision grounds for allocation to individual slaves, they generally showed little interest in the ways that the enslaved managed food production in these remote sites. It was on such distant provision grounds that slaves later in the eighteenth century broadly won the "right to grow their own food crops," thereby extending the practice of independent production to islands where it previously had no influence.[33]

On sugar islands of low topographic relief, food shortages grew acute as cane cultivation overtook most of the arable land. Planters responded by importing food for their workforce. A regional trade in food soon developed. Mainland North America became an important supplier, especially of salted fish, salt beef, grain, and legumes.[34] Carolina planter Henry Laurens found markets in the British Caribbean for the rice and the black-eyed peas that his enslaved growers produced.[35] Subsistence shortages in other sugar-producing areas, such as eighteenth-century Danish St. Thomas, Antigua, and Dutch Guiana, similarly induced some plantations to specialize in the cultivation of food for others. The semiarid island of Barbuda, for example, supplied livestock and crops to Antiguan sugar estates, which lay thirty miles from its shores. Among the crops produced by Barbuda's slave community were "yams, 'pease,' and corn."[36]

Slaves on provision plantations thus grew the foodstaples that fed other enslaved workers on food-deficit sugar islands. Both maize and guinea corn were marketed as food rations in the intraregional trade. In the Guianas, provision estates supplied the food-deficit coastal sugar plantations with plantains, eddoes, and pigeon peas.[37]

This regional provision trade was slave-based and oriented to the production of Amerindian and African subsistence staples—the tropical grains, tubers, and legumes that each cultural legacy had developed for food security. In this food-import economy, some African crops that the first generations of enslaved Africans pioneered in their food fields were fed to their descendants in the form of mass-import rations, as occurred with guinea corn on Barbados and Curaçao.

Despite these regional adjustments in food supplies, the substitution of imported rations for locally produced food narrowed the range of dietary staples available to slaves who toiled on monocrop sugar islands. A food-ration system resulted in an unvarying and nutritionally inadequate diet. Moreover, the physical spaces available for slaves to improve and diversify their diet contracted to just the small yards surrounding their dwellings.[38] Vital food options, and the means to express them, disappeared.

The fragile mechanism of food-import dependency broke down during the second half of the eighteenth century. The tumult of serial international conflicts—the Seven Years' War, the American and French revolutions, and the Napoleonic Wars that followed—severely disrupted the regional food trade. Recurrent subsistence shortfalls resulted in hunger on some islands, outright starvation on others. Food scarcity was in fact a contributing cause of the unrest that sparked the Haitian Revolution in 1791.[39] The revolutionary upheavals and disruptions in the Caribbean in the remaining decades of the eighteenth century forced import-dependent islands to return to self-sufficiency as a matter of survival. Land once given to sugar was again placed in provision grounds.

On some of these newly established provision grounds, supervised labor gangs planted the food crops. But this system coexisted with, and was often replaced by, one that increasingly resembled the Pernambuco system, in which slaves were allocated individual plots to grow their own food.[40] The practice of independent production grew throughout the Caribbean in the final decades of the eighteenth century. As a consequence, the average size of the food plot allotment in the British West Indies nearly doubled in the fifty-year period 1750–1800, from four-tenths of an acre per slave to seven-tenths of an acre.[41]

INDEPENDENT PRODUCTION
IN THE AFRICAN-ATLANTIC WORLD

The convention of independent food production was an early feature of the African-Atlantic world. European observers of African societies describe it during the initial period of the transatlantic slave trade. Historian John Thornton discusses its existence in São Tomé, a sugar-producing island off the coast of present-day Equatorial Guinea, through the words of an anonymous Portuguese pilot who visited five times between the 1520s and 1540s. The "rich men of São Tomé had large groups of slaves ranging from 150 to 300 who had the 'obligation to work for their master every day of the week except Sunday, when they worked to support themselves.'.... The masters 'gave nothing to the said blacks,' neither food nor even clothing, which they had to make for themselves from local products in their own time." The practice drew the concern of Carmelite missionaries on the island in the 1580s, since slaves were given the Sabbath to work for their own subsistence even as they were relieved of their plantation labors on that day. The Carmelites expressed concern that slaves were working "on feast days and Sundays to produce for themselves, which they argued should be days of rest."[42] The Carmelites apparently did not comprehend a practice in which servitude also vested slaves with the right to sufficient time to attend to their own subsistence.

Independent production did not arise sui generis in New World plantations. Its diffusion probably owes as much to conventions of indigenous African slavery as planter strategies to gain economic advantage. Independent production was, for instance, evident during the commercialization of millet cultivation in slave-trading Senegambia.[43] The early sixteenth-century chronicler Valentim Fernandes recorded that Wolof slaves worked one day a week for themselves. European commentaries on Senegambia centuries later describe the long-standing tradition in indigenous African slavery whereby slaves typically retained one or two weekdays for their own food production. On the days worked for their master, the practice was to labor from sunrise to midafternoon. The practice of leaving slaves to generate their own food recognized the need to allocate time for its production.

Some slaves held in indigenous African slavery were able to obtain an individual food plot and the time to work it if they distinguished themselves in service to their masters. For those able to achieve this level of trust, independent production conferred a measure of protection against the risk of sale to European slave traders. Such concessions were apparently negotiated individually but were not an obligation of the master.[44] The convention of placing slaves in charge of their own subsistence by the grant of independent plots reappeared in the sugar-producing estates of tropical Brazil, but stripped of the protections and rights conveyed in its African formulation. Plantation owners knew little about growing food in the tropics, so putting subsistence production in the hands of their bondsmen represented a practical, if not convenient, solution. What appears as a convention bound by rules and tradition in indigenous African slavery had become an unfettered instrument of economic expediency on New World plantations. Any advantage an enslaved plantation worker may have gained by access to independent food plots was thoroughly subverted to the year-round demands of sugarcane. Altered by the relentless commercial priorities of European slaveholders, the time allotted individual slaves to work their subsistence plots was scarcely a priority.

There is a considerable scholarship on independent production, which discusses the origins and geographical and temporal expansion of the convention throughout plantation societies of the New World tropics. Much has been written on independent production as an economic strategy, with emphasis on the surplus food that slaves marketed and the plots' role in reducing slaveholder expenses. From the planters' perspective, independent production enabled them to reduce food allotments to the enslaved workforce or to forgo responsibility altogether. Nevertheless, slaves repeatedly fought for the right to a subsistence plot even when its recognition meant an intensification of their own labor burden.[45] This is because the plot represented considerably more than the physical space for growing food. Access to land for independent production gave slaves the opportunity to plant crops without supervision, to produce beyond their dietary needs, to realize petty cash or goods from marketable surpluses, and to derive direct benefits from a portion of their overall labors. Independent production provided both a degree of economic autonomy and an opportunity to strengthen subsistence security. But it was also the genesis of attempts to extract broader acknowledgment that not all of one's labor was owned by the master. The significance of the institution was the burgeoning notion that an enslaved person had the right to some of the time he or she labored and to the products that came from this work.

The convention of independent production expanded across Caribbean plantation societies over the second half of the eighteenth century following the food shortages induced by the upheavals of European conflicts. These subsistence plots represented the physical sites where slaveholders acknowledged that a portion of plantation labor time and crops belonged to their slaves. Once established, the convention of independent production sometimes served as a springboard for negotiating additional rights, such as the ability to bequeath the subsistence plot to one's family or a person of one's choosing, as Woodville Marshall writes of the Windward Islands: "Slave families in the 'constant occupation' of provision ground forced their owners to recognize rights of occupancy to portions of plantation ground. Slaves would not move from their [provision] ground without notice or without replacement grounds being provided, and they could bequeath rights of occupancy as well as property."[46]

A similar effort was reported (ca. 1815) on a sugar plantation in Northeast Brazil operated by Benedictine monks. "The slaves are allowed the Saturday of every week to provide for their own subsistence, besides the Sundays and holidays . . . and when a negro dies or obtains his freedom, he is permitted to bequeath his plot of land to any of his companions whom he may please to favour in this manner."[47]

The right to designate the subsistence plot's heirs was also reported in the French Caribbean on some of the islands visited by French abolitionist Victor

Schoelcher during 1840–42. Just years before the French Republic's emancipation decree (1848), Schoelcher noted that slaves worked their subsistence plots "communally," that is, with the labor of family members and other kin. He also reported that they held established rights that slaveholders were compelled to recognize, which included the right to leave the subsistence plot and its produce to their relatives or designated heirs. "They pass them on," wrote Schoelcher, "from father to son, from mother to daughter, and, if they do not have any children, they bequeath them to their nearest kin or even their friends."[48] From such roots in independent production, the tradition of family land took hold in many parts of the Caribbean.[49]

The diffusion of independent production as the predominant subsistence strategy in the late eighteenth-century Caribbean brought new challenges to enslaved people. Foodstaple cultivation frequently took place in marginal environments, on soils affected by poor drainage, acidity, and stoniness or in degraded areas such as eroded ravines and mountain slopes. These subsistence grounds often required rehabilitation before they could be coaxed into production. They also necessitated crops that did not require constant attention. Under such subsistence challenges slaves brought to bear the full range of their experience as tropical farmers to make the diverse cultivation environments yield. They stabilized soil banks from erosion by ridging the slope with hoes. They selected dietary staples for ease of cultivation and productivity. Root crops such as plantains, yams, and taro were especially favored. So were legumes such as the tropical pigeon pea, which was appreciated as food but also for improving soil fertility, as provender for domestic animals, and as a plot boundary marker.[50]

Crops requiring additional care and inputs—typically vegetables, herbs, spices, medicinals—were grown in the dooryard gardens outside slave dwellings, as George Pinckard reported from Barbados in 1796: "On these small patches of garden it is common for the slaves to plant fruits and vegetables, and to raise stock. Some of them keep a pig, some a goat, some guinea fowls, ducks, chickens, pigeons, or the like." In these locations the work of women and the aged was especially evident. Food wastes and manure from small animals penned in the yard improved the garden's productivity.[51]

The changing face of subsistence cultivation on sugar islands over the eighteenth century—from food self-reliance to imports and back again—occurred against the backdrop of the ongoing disembarkation of enslaved Africans to the Caribbean. The new arrivals repeatedly injected African agricultural practices and methods, with important consequences for import-dependent islands where food cultivation had contracted to the yard surrounding slave dwellings. They also brought with them knowledge or experience of the conventions common to indigenous slavery practiced in Africa, conventions stripped away in transatlantic slavery, but experienced as rights by its bondsmen. The steady infusion of enslaved Africans, and the crops and practices descendants of earlier migrations had maintained in kitchen gardens, facilitated the return to food self-sufficiency in the region. This was achieved with the expansion of subsistence cultivation beyond the household plots to provision grounds and independent plots. In such ways, Africans continually shaped the food systems of tropical plantation economies.

BOTANICAL GARDENS OF THE DISPOSSESSED

The many novel plants that European plantation owners, visitors, and naturalists described in the food fields of slaves enumerate species that we now know are of African origin or were an established foodstaple on the continent during the Atlantic slave trade. Included among the African introductions that Jamaican slaves planted in their food fields were the African oil palm, yam, pigeon pea, banana and plantain, kola, pearl millet, guinea corn, rice, taro, the bottleneck gourd, and okra. The African components of subsistence staples in other plantation societies included hibiscus (known as *jamaica* in the Spanish Caribbean), black-eyed and lablab peas, the Bambara groundnut, sesame, guinea squash, guinea pepper, melegueta pepper, and jute mallow.[52] Such an extraordinary range of African crops made "the slave provision ground," in the words of one eighteenth-century observer of Saint Domingue, "*une petite Guinée.*"[53]

A sampling of the African introductions found in these subsistence sites—as reported by slaveholders, plantation observers, naturalists, and modern scholars—is presented in figure 7.1. The fundamental necessity of food to human life provides the context for understanding the subsistence strategies slaves developed in plantation societies. As European commentaries repeatedly indicate, the African botanical introductions initially gained their New World footing in the food plots of enslaved Africans. In these small and fragmented spaces of food production, Africans realized an alternative botanical vision to the plantation export commodities that were vested with the dehumanizing practices of the plantocracy. Here, slaves organized cultivation for their own purposes, selecting plants that improved their diet, healed their bodies, and provided them spiritual succor in the liturgical practices of Africa-based religions. As informal experimental stations for the transfer, establishment, and adaptation of African food crops and dietary preferences, these plots became the botanical gardens of the Atlantic world's dispossessed. The apotheosis of this subaltern experience is exemplified in the story of one descendant, George Washington Carver

FIGURE 7.1. African plants established in the plantation era.

CEREALS

Millet	*Pennisetum glaucum*	B, Ca, H, Hi, J, L, La, Po, S, We
Sorghum	*Sorghum bicolor*	Ca, H, Ha, Hg, Hi, Hr, K, L, La, Ma, Pu, S
Rice	*Oryza* spp.	H, Hg, Hi, Ho, Hu, Mc, Pr, S, Sx, Wl, Wo

TUBERS

Yam	*Dioscorea cayenensis*	Ca, H, Ha, Hg, Hi, Hr, O, OM, Pr, Pu, S, St, Wa
Plantain/banana	*Musa* spp.	H, Ha, Hg, Hm, Hr, K, Ku, Li, O, OM, Pi, Pr, Pu, S, St, Wa
Taro/eddo	*Colocasia esculenta*	H, He, Hr, K, Wa

LEGUMES

Black-eyed pea/ cowpea/calavance	*Vigna unguiculata*	K, L, S, Sp, Wl
Pigeon pea/Congo pea/Angola pea/ guandul	*Cajanus cajan*	H, Hg, Hi, Hr, Pi, Pr, Pu, St, Sx
Bambara groundnut/ Voandzeia	*Vigna subterranea*	M, Pr
Lablab/bonavist/ hyacinth bean	*Lablab purpureus*	Hg, Hr, Je, K, L, Li, M, Pi

BEVERAGES

Coffee	*Coffea* spp.	Hr, Pr, Pu, St
Tamarind	*Tamarindus indica*	H, Hi, Hm, Hr, Li, M, Pi, Pu, St
Roselle/hibiscus/ bissap	*Hibiscus sabdariffa*	H
Kola nut	*Cola* spp.	E, K, P, S, V, Wl

Historical observer(s) or modern scholar(s), location:

B = Jean Barbot, French Guiana
C = Lydia Cabrera, Cuba
Ca = Mark Catesby, the Carolinas
Cs = Luigi Castiglioni, Carolina
D = Jean Baptiste Debret, Brazil
E = Bryan Edwards, Jamaica
G = Lewis Gray, South Carolina
H = Frederick Hall, William Harrison, and Dorothy Winters Welker, Northeast Brazil
Ha = Jerome Handler, Barbados
He = Karen Hess, Virginia

Hg = Barry Higman, Caribbean
Hi = Arnold Highfield, Danish West Indies
Hm = Alexander von Humboldt, Colombia
Ho = F. C. Hoehne, Brazil
Hr = David Harris, outer Leeward Islands
Hu = J. Hurault, French Guiana
J = Han Jordaan, Curaçao
Je = Thomas Jefferson, Virginia
K = Clarissa Thérèse Kimber, Martinique
L = Jean-Baptiste Labat, Martinique and Guadelope
La = John Lawson, Carolina

OIL PLANTS AND FRUITS

Sesame/benne	*Sesamum radiatum*	Hu, Je, M, P, Pi, S, W, Wl
Castor bean	*Ricinus communis*	C, Hr, K, M, Mc, S, V
Oil palm/*dendê*	*Elaeis guineensis*	H, S, V, Wl
Watermelon	*Citrullus lanatus*	Hi, Hr, Hu, P, Pi, St, Sx
Muskmelon	*Cucumis melo*	K, La, Wo
Ackee	*Blighia sapida*	E, P, Ru

VEGETABLES AND SPICES

Okra	*Abelmoschus esculentus*	Cs, He, Hi, Hr, Hu, K, M, P, Pi, Pr, S, St, Sx
Vegetable amaranth/ African spinach/ bledo/callalou; jute mallow/ bush okra	*Amaranthus* spp.; *Corchorus olitorius; Vernonia* spp.	C, Hi, V, Wl
Melegueta pepper	*Aframomum melegueta*	H, He, Ho, K, M, V
Guinea pepper	*Xylopia aethiopica*	C, Ru
Guinea squash	*Solanum aethiopicum*	H, He, Ho, Hr, La, Pi

UTILITY

Bottleneck gourd	*Lagenaria siceraria*	H, M, V
Kenaf	*Hibiscus cannabinus*	H

FODDER

Guinea grass	*Panicum maximum*	C, D, Hr, K, P, S, Sx, Wo
Pará/Angola grass	*Panicum muticum*	D, Hr, K, R, Wa
Bermuda grass	*Cynodon dactylon*	G, K

Li	= Richard Ligon, Barbados	S	= Hans Sloane, Jamaica
M	= Georg Marcgraf, Pernambuco, Brazil	Sp	= Randy Sparks, South Carolina
		St	= John Gabriel Stedman, Suriname
Ma	= Woodville Marshall, Windward Islands	Sx	= Johann Baptist von Spix and Carl Friedrich Philipp von Martius, Brazil
Mc	= James McClellan, Saint Domingue		
O	= Gonzalo Fernández de Oviedo, Spanish Caribbean	V	= Robert Voeks, Bahia
		Wa	= David Watts, Barbados
OM	= John Oldmixon, Barbados	W	= Mary Tolford Wilsón, South Carolina
P	= John Parry, Barbados and Jamaica		
Pi	= Willem Piso, Brazil	We	= Waldemar Westergaard, Danish West Indies
Po	= Johannes Postma, Curaçao		
Pr	= Richard Price, Suriname	Wl	= Maureen Warner-Lewis, Trinidad
Pu	= Lydia Pulsipher, Montserrat	Wo	= Peter Wood, South Carolina
R	= Juan Tomás Roig, Cuba		
Ru	= A.M.G. Rutten, Curaçao		

(ca. 1864–1943). Born in slavery in Missouri, Carver gained scientific renown through his work on three seemingly minor crops—okra, the black-eyed pea, and the peanut—each long associated with the African presence in mainland North America and a staple of slave food gardens.

A critical feature of human migration the world over is the preservation of traditional dietary preferences across space and the dislocations of geography. That the migration of Africans was compelled through extremes of violence and cruelty does not diminish this universal desire or preclude the possibility of achieving it. African staples enabled slaves at times to reinstate some food traditions of specific cultural heritages and to combine ingredients in new ways with Amerindian and European foods. In this way, slaves discretely modified the monotony of any food regimen slaveholders might impose. The introduced African crops encouraged the distinctive foodways that eventually developed across plantation societies. Africans and their descendants thus profoundly shaped the culinary traditions of slave societies, combining in new ways the foods of three continents in their struggle to secure daily sustenance. They moreover realized this achievement under circumstances no other immigrant group had to face. Africa's botanical legacy in the Americas is built upon this unacknowledged foundation.

EIGHT

Guinea's Plants and European Empire

A *Portuguese* vessel arrived, with slaves from the east, with a considerable quantity of rice, being the ship's provision: this rice the *Carolinians* gladly took in exchange for a supply of their own produce.

A. S. SALLEY
"Introduction of Rice into South Carolina"

EUROPEANS IN THE NEW WORLD tropics discovered many plants new to them in the food plots of slaves. Having often no names for these plants and lacking an appropriate classification or nomenclature, they simply borrowed the African-language names from the slaves who grew them. Some of these loan words were absorbed into the borrowers' tongues and survive to this day as the popular word for the crop. This process of applying extant vocabulary to new knowledge was especially vigorous in the early plantation period, as peoples and plants from three continents were thrown together.

When Gonzalo Fernández de Oviedo reported Hispaniola's slaves growing and eating a newly introduced tuber in the 1520s, he adopted the African name they gave it and recorded it as *nname. Nname* (or *ñame* in Spanish) is actually the African yellow yam.[1] The word *ñame* derives from the Portuguese rendering, *inhame*, of the Mandinka word *nyambi*, with *nyam* meaning "food" or "eating" in several western African languages.[2] The common names for yam in Spanish and Portuguese (*ñame de Guiné* and *inhame da Guiné*, respectively) embrace both a toponym and a vernacular name. Yams became a major foodstaple of the transatlantic slave trade from its earliest period. At this time, the New World potato (*Solanum* spp.) had not yet been adopted by Europeans, and they had never seen or tasted anything like the yam, being for the most part unacquainted with the class of edible starchy foodstaples grown from tubers and not from seed.[3] Thus, it is not surprising that finding no equivalent for the yam, they took a western African word for it. Ironically, the vernacular term is a more accurate descriptor of the yellow yam's origin. Its scientific name, *Dioscorea cayenensis*, incorrectly suggests Cayenne (French Guiana) as its geographical origin.

Georg Marcgraf, on his seventeenth-century scientific expedition to Dutch Brazil, routinely applied place names and vernaculars to new plants he found there. For most species he records the indigenous Amerindian name; for others he provides an African-language synonym. The use of these appellations is interesting, not because we can vouchsafe his taxonomic accuracy (which we often cannot), but because Marcgraf consistently cedes the authority for naming a new plant to African or Amerindian sources. The practice of borrowing their words for novel foodstaples underscores European unfamiliarity with these tropical plants and defers to the indigenous knowledge of Amerindians and Africans.

Europeans adopted African names for many introduced plants they discovered in the subsistence fields of their slaves. Okra is an African domesticate and a popular vegetable and soup thickener of diaspora cuisines. Words similar to it transferred from Akan and Igbo languages of the Lower Guinea Coast to several colonial languages of the New World.[4] Suriname's Maroons call it *oko*. The African vegetable is also known by various Bantu-language derivations as *gumbo* (in both English and Portuguese), and *quiabo* or *quingombo* (Portuguese).[5]

Sesame was usually present among the crops Europeans identified as growing in slave subsistence plots. In Dutch Brazil, Marcgraf described and drew it, noting sesame had been introduced from Africa. He recorded the plant's Congo name, *gangila*, likely the source of the Portuguese word for it, *gergelim*.[6] On the introduction of sesame to South Carolina, Thomas Jefferson claimed it was "brought from Africa many years ago by their negroes & by them called Benney."[7] Sesame was often known in the English colonies by its West African vernacular names: *bene, benne,* and *beni* (in the Wolof, Mandinka, and Bambara languages, among others).[8] Plantation owners first found it in slave subsistence plots and observed the ways it was used. In the 1770s Bernard Romans reported that "the Negroes use it as food either raw, toasted, or boiled in their soups and are very fond of it."[9] Thomas Jefferson added that they "bake it in their bread, boil it with greens, [and] enrich their broth [with it]."[10] The seed was pressed for cooking oil and the leaves used as greens; when added as a potherb to soups or one-pot stews, sesame leaves thicken the broth. When planters learned to exploit sesame's commercial potential, its transmutation from botanical curiosity to profitable commodity was completed. As Europeans first borrowed the vernacular names for new plants from Africa, so did they eventually come to appropriate the plants and knowledge to grow them.

The pigeon pea bears its Bantu names in many New World languages. It is *gungo* in Jamaica; *gandule* in some parts of the English Caribbean; *wando* in the Dutch Caribbean; *guandu* or *andu* in Portuguese Brazil; and *guandul* in Spanish America.[11] The African kola nut is known as *bissy* or *bichy* (from the Akan and Gbe languages) in Jamaica and the Gullah spoken on the Sea Islands off South Carolina and Georgia; it is *goora* in Martinique (from Hausa and Fula) and *obi* or *orobô* (Yoruba) in Afro-Brazilian religions.[12] The African oil palm, which gives Bahian cooking its defining taste and color, is *dendê* in Brazil (from various Bantu languages) and *abbay* in Jamaica, from the Twi language.[13] The Asian taro, which arrived in the New World from Africa, is called by its Fante name *eddo* in Jamaica.[14] Each of these plant names derive from west and west-central African languages. Even the name of the world's most popular fruit, the banana, is African in origin.[15]

The Bambara groundnut *(Vigna subterranea)* is an African legume, high in protein, which found its way to subsistence plots in tropical America. Described in Brazil in the 1640s and a century later in Suriname, it was undoubtedly familiar provender aboard slave ships.[16] The seventeenth-century Dutch factor Willem Bosman, operating out of the Gold Coast fort of Elmina, encountered Bambara groundnuts, which he called "*Angola* Beans," among several types of earth nuts that were cultivated there.[17] On the other side of the Atlantic, Marcgraf described the same plant in Dutch Brazil, calling it *mandubi d'Angola*—the Angolan peanut. His nomenclature attaches the Amerindian Tupi-language word for the Amerindian peanut *(mandubi)* to a place name that identifies the introduced groundnut as being from Angola.[18] More than a century later, Moravian missionaries in Suriname reported the Saramaka Maroons planting the Bambara groundnut, which the Saramakas called *gobogobo*. The word is one of several related names from western Africa for the indigenous groundnut.[19]

The Bambara groundnut shares a curious history with the peanut, its New World counterpart. Amerindians domesticated the peanut thousands of years ago in tropical eastern Bolivia. By the time of Columbus, cultivation had reached some parts of the Caribbean but not yet Central or North America. The Arawak Indians of the Antilles called it by the Tupi cognate *maní*, a name that was adopted by Spanish colonizers and persists today on some Spanish-speaking islands. Portuguese caravels carried the peanut from Brazil—where the Portuguese first called it by its Tupi name *mandubi*, later modified to the modern Portuguese *amendoim*—to West Africa in the sixteenth century. By the 1560s the peanut was widely cultivated in Guinea. Africans adopted it into their agricultural systems, as they had maize and other Amerindian foods. But Africans did not borrow the New World names for the peanut. Instead, they applied variations of *gobogobo* and other vernacular names for the familiar

Bambara groundnut. Even Bosman borrowed the western African vernacular name for the peanut: "Here is also another sort called *Gobbe-gobbes,* which grow two together in a Cod under the Earth, and shoot out a small Leaf above the surface of the Earth."[20] Eventually, the African-language names *nguba* and *mpinda* passed into English as the vernacular terms for the peanut, "goober" or "pindar."[21]

Protected by its shell, the peanut was ideal transatlantic provender, able to survive extended sea voyages with minimal spoilage. So began the peanut's second Atlantic crossing—this time to mainland North America—aboard slave ships departing Africa. As the peanut could be eaten either cooked or raw, it quickly became a versatile staple of the Middle Passage. The nut arrived on so many slave ships that many English commentators thought it an African plant. Hans Sloane mentioned the peanut's importance as provision on seventeenth-century slave ships to Jamaica even though it was not yet known by that name: "The Fruit, which are call'd by Seamen Earth-Nuts, are brought from *Guinea* in the *Negroes* Ships, to feed the *Negroes* withal in their Voyage from *Guinea* to *Jamaica.*"[22] The plant was a novelty to Sloane, as it was to his contemporary and correspondent, naturalist Henry Barham. Arriving in Jamaica near the end of the seventeenth century, Barham wrote of *pindalls:* "The first I ever saw of these growing was in a negro's plantation, who affirmed, that they grew in great plenty in their country; and they now grow very well in Jamaica. Some call them *gub-a-gubs;* and others ground-nuts, because the nut of them, or the fruit that is to be eaten, grows in the ground.... They may be eaten raw, roasted, or boiled. The oil drawn from them by expression is as good as oil of almonds."[23]

Barham discovered the peanut in the food fields of an enslaved African from whom he seems to have acquired his information. He uses two African names for it, describes how it was consumed, and commends the quality of the extracted oil. The peanut and its useful properties were certainly not a novelty to some of Jamaica's slaves, the plant having been a de facto crop in West Africa for over a century. In fact, one of the earliest areas settled by Maroons in Jamaica bears the toponym "Pindars."[24]

Even though the peanut was widely grown as a subsistence staple in slave food plots, whites did not initially accept it as a foodstaple or value its pressed oil. But in the plantation economies of mainland North America they did appreciate the oilseed "cake" by-product as an excellent feed for swine and poultry.[25] In a pattern that would characterize several other crops found in slave food fields, mercantile attention to the peanut began in the decade prior to the Revolutionary War. Initial interest in it was as a source of vegetable oil rather than as a confectionary (Americans and Europeans did not acquire a commercial taste for the peanut until the end of the eighteenth century).[26] In 1769 the Royal Society received a sample of peanuts and peanut oil from George Brownrigg, a North Carolina plantation owner, for consideration as a substitute for the metropole's dependence on imported olive and almond oil.[27] In published correspondence endorsing Brownrigg's idea, one society member described how the peanut came to the English colonies from Africa and owed its dissemination to slaves:

> I lay before you some pods of a vegetable, and the oil pressed from their contents.... [They] are the produce of a plant well known, and much cultivated, in the Southern colonies, and in our American sugar islands, where they are called ground nuts, or ground pease. They are originally, it is presumed, of the growth of Africa, and brought from thence by the negroes, who use them as food, both raw and roasted, and are very fond of them. They are therefore cultivated by them in the little parcels of land set apart for their use by their masters. By these means, this plant has extended itself.[28]

Plantation owners such as Brownrigg first encountered crops like the peanut in the food plots of slaves, from whom they learned about their culinary uses. With this crucial knowledge, they explored and then developed the crop's commercial potential. In this way, planters comfortably straddled two important agricultural domains: the subsistence domain of their slaves, in which they discovered many new crops, and the commercial domain of commodity production. It is perhaps not accidental that it is a plantation owner who first recognizes the value of the indigenous knowledge held by his slaves, and catapults it into the realm of commerce.[29]

For enslaved Africans the peanut provided an important source of protein to a diet monotonously dominated by starches. Expertise with the Bambara groundnut had encouraged African experimentation with the peanut when it was brought to Africa in the sixteenth century.[30] It also informed the ways slaves consumed the peanut—raw, boiled, roasted, and in stews combined with other foods.[31] Just as the English in seventeenth-century Ireland scorned the New World potato as "Irish food," or plantation owners in the Caribbean the plantain, the peanut was likewise initially repudiated as "slave food" in the South.[32]

The social and racial prejudices that divided slaveholders from those they enslaved also kept separate the foods they ate. Planters did not initially consume the staples associated with their slaves. But this wall of culinary segregation gradually disintegrated over the centuries as signature ingredients of the African diaspora stealthily made their way into white kitchens and onto white tables.[33]

African foodstaples infiltrated the cuisine of slaveholders through the dishes and confections enslaved cooks prepared for them. Today, raw and boiled peanuts and peanut brittle are considered heritage food in the South, as are sorghum molasses, okra-based gumbo, Hoppin' John (rice and black-eyed peas), and other dishes based on crops that slaves pioneered in their food fields. Each testifies to the important ways that slaves shaped subsistence and foodways in former plantation societies. The food traditions of these societies today provide a long overdue opportunity for black and white alike to discuss a heritage now fully integrated, but whose African contribution is still not yet fully appreciated.

AFRICAN MILLETS IN THE NEW WORLD TROPICS

African knowledge was easily used against the slaves who held it. This can be seen in the ways that guinea corn was imposed as the dietary staple on several Caribbean islands. The quest for a viable export staple in some seventeenth-century colonies prompted plantation owners to take interest in the crops their slaves grew. They experimented with, and often realized, the commercial potential of several introduced African botanical species.

Africans in ancient times developed the world's two most drought-tolerant cereals, sorghum and pearl millet. While millet will grow where no other cereal can, sorghum is a surpassing botanical marvel because it can be planted in both temperate and tropical zones, with or without much rainfall. It grows in infertile soils, tolerates salinity, and produces a cereal that minimally taxes soil nutrients. Africans traditionally intercrop sorghum with the indigenous nitrogen-fixing legumes (cowpeas, pigeon peas) that rejuvenate exhausted fields. For these reasons, sorghum or guinea corn (as it was known during the slave trade), is a more reliable crop than maize in dry regions.[34] Sorghum has the added benefit of being an excellent animal feed.[35]

When the African "millets" appeared in plantation economies, Europeans repeatedly credited slaves with their introduction, even if they believed the cereals were originally native to India. The eighteenth-century naturalist Charles Bryant summarized prevailing views: "[Sorghum and millet are] . . . native[s] of India. . . . Both these plants are cultivated in Africa by the name of Guinea Corn, and they have been confounded as only one sort by most travellers. The grain is there made into bread, and otherwise used, and is deemed wholesome food. From Africa the Negroes carried them to the West Indies."[36]

In the early colonial period, millet and sorghum were commonly mentioned as plants that slaves preferred and grew in their food fields.[37] As early as 1587,

Gabriel Soares de Sousa noted in Bahia, Brazil, that African slaves favored planting guinea corn for food over New World maize.[38] A century later, Hans Sloane found that slaves planted sorghum "every where in *Jamaica* for Provision, yielding very great increase. . . . Cakes are made of its Flour . . . it is sown at a Foot distance, three or four Grains into a hole."[39] He was among the first European observers to distinguish sorghum from millet, noting that the latter "has lesser Grains or Seeds than the former. . . . It is to be met with in some Negro's Plantations, though not so commonly as the former [sorghum]."[40] A dozen years later, John Lawson similarly observed slaves planting guinea corn in their subsistence plots in the Carolina colony.[41]

But the role of the African millets was far more pronounced in the Caribbean. Many islands in the New World tropics are, contrary to the popular perception, actually quite arid, characterized by parched landscapes and scant rainfall.[42] In other areas precipitation is sparse because of the island's location in the rain shadow of tropical storms. The climate of the Bahamas, Curaçao, Antigua, Barbuda, Anguilla, and the leeward side of other islands in fact resembles the low and erratic precipitation regimes of the African Sahel. Millet and sorghum were well adapted to these dry environments. They were to the semiarid tropics what yams and plantains were to humid areas. Each grew where moisture-loving maize would not. Their advantage over Amerindian manioc as a subsistence staple was the relative ease of processing. As Caribbean islands were transformed into plantation societies, the role of the African cereals was recast.

On the drought-prone island of Barbuda, African millets were grown to supply food-deficit sugar plantations. In 1685 the English Codrington brothers were granted a royal lease to the island, which their descendants held for nearly two centuries thereafter. They used Barbuda to raise animals, meat, and victuals for sale on nearby Antigua and to supply the family's sugar estates there. The entire island was converted into a pasture, where enslaved Africans tended cattle, sheep, goats, horses, and donkeys for their masters. Documents in the early eighteenth century indicate that slaves managed two grazing and cultivation systems, one for export to Antigua, the other for their subsistence. The provisions grown were the same in both systems. By 1715 slaves were raising yams, "pease," and corn. "Corn" referred to both maize and guinea corn, and "pease" likely included both Amerindian beans and the African pigeon pea.[43] Grazing and cultivation on Barbuda were not isolated activities, but complementary sides of an African land-use system, in which slaves seasonally shifted cultivation sites to pastureland that was manured. In both crop choice and animal husbandry practices on drought-prone Barbuda, African influences are evident.

Another Caribbean island that became dependent on the African millets for subsistence was Curaçao. When the Dutch established the colony in the 1630s, its principal virtue was the island's excellent deep-water harbor. The dry climate precluded sugar production, and by the 1660s Curaçao had become a holding facility for slaves destined to be sold on the South American mainland, just forty miles away. Slaves retained on the island worked as dockhands, on the island's salt pans, and in fishing. They also tended livestock and cultivated foodstaples. As a major slave depot in the transatlantic trade, Curaçao required considerable reserves of food. African millets, easily adapted to the semiarid conditions of the island, kept slaves from starvation.

Even so, as Dutch involvement in the transatlantic slave trade escalated, food shortages became a recurrent problem. In 1687 the Dutch West Indies Company (WIC) cancelled four scheduled slave shipments to Curaçao because of insufficient food reserves. WIC officials responded by dispatching a delivery of food from West Africa to forestall the starvation of the slaves crowded into the island's detention centers.[44] It is in the context of the subsistence crises of the 1680s that the importance of sorghum as the leading foodstaple first comes to light. The need for food was so acute on Curaçao in the years prior to 1700 that manumission reputedly could be granted if a slave's sorghum production exceeded three hundred bushels.[45] By 1725, the island's arable land was entirely given to the cultivation of "guinea grain." Sorghum and millet were used to feed the swelling population of new arrivals and the livestock raised on other parts of the island.[46]

Sorghum became so central to Curaçao's food supplies that Governor Kikker detailed in 1817—nearly a half century before slavery was abolished—how slaves cultivated it. Sown in October and harvested about six months later, sorghum was planted with methods identical to those used in West Africa. Slaves carried the seeds in the African bottleneck gourd to the fields. A hole was made in the ground into which the seeds were dropped; a sweep of the foot across the soil covered the planted area. At harvest time slaves protected the exposed seed clusters by scaring away flocks of birds, which would gather to devour the ripening fields.[47]

The sorghum cultivation cycle concluded with a harvest festival that involved the entire island population.[48] A Dutch newspaper article published in 1838 underscored the historic importance of sorghum to the settlement of Curaçao: "Among the crops with which the Creator has blessed the tropics, none can surpass the guinea corn. It appears in places, where because of the drought no other plants can grow. Much of the population on our island and other tropical places, have to show their thanks to this crop. This crop fulfills the same role that potatoes do for Europe."[49]

Curaçao's celebration of the sorghum harvest offered in its own way a trop-ical version of Thanksgiving, but one the enslaved commemorated with an in-digenous African crop. The long struggle with hunger and food availability on the island provides context for understanding the significance of the harvest for its populace.[50]

While sorghum proved a viable food source in the dry tropics, both sorghum and millet played an important role on Caribbean islands that received ample rainfall—Barbados, Jamaica, the Danish Virgin Islands, and Haiti. A Moravian missionary to the Virgin Islands in the 1760s provided an accurate description of millet when he described the importance that "small magis" (the African mil-lets in contrast to larger-grained maize) had assumed in plantation food fields:

> Small Magis [millet] grows much taller than the former variety [maize]. However, it has small round kernals which are the size of hemp-seeds, arranged on a long cylindrical cob. Moreover, it is a useful and wholesome fruit for both men and beasts. It is an especially good food for the slaves, who do hard work. They get along just as well with the cooking of the small variety as they do with the large. Large quantities of it are planted in order to have ample provisions, as it can be stored for a long time.... In addition to these two, still another kind of small magis is raised there, which is called Guinea corn.[51]

As in the Virgin Islands, guinea corn in eighteenth-century Jamaica was "a prin-cipal part of the food of the Negroes."[52]

Many of these islands were planted in sugarcane. On Barbados, sugar culti-vation quickly commanded the available arable land, and this weakened the island's ability to feed itself. African crops were part of the solution. Sorghum's ability to grow in exhausted sugar soils served as the fulcrum of a strategy to re-plenish the fertility of existing fields and improve subsistence availability. This strategy was uniquely African in origin: instead of leaving the land fallow, sorghum was interplanted with African legumes (typically pigeon peas or the lablab bean). The nitrogen-fixing legumes renewed soil fertility, and the unde-manding sorghum provided both a human and animal food source.[53]

Guinea corn was rarely eaten by planters and so was naturally cast as food for slaves. Depleted sugar soils in Barbados meant that occasionally more planta-tion ground was planted in sorghum than sugarcane.[54] Guinea corn's tolerance of nutrient-deficient soils gave slaveholders new means to achieve subsistence on sugar islands where arable land was scarce. Sorghum, part of a traditionally diversified food complex in Africa, was made the principal slave subsistence ration on many mature sugar islands. In 1783 Charles Bryant reported that in the British West Indies "each slave is generally allowed from a pint to a quart [of sorghum] per day."[55] The monotonous diet of guinea corn led a visitor to Barbados to record one enslaved man's complaint of its unvarying use as a subsistence staple: "Massa gib me Guinea corn too much—Guinea corn to-day—Guinea corn to-morrow—Guinea corn eb'ry day—we no like him Guinea corn—him Guinea corn no good for gnhyaam [eat/*nyam*]." But on Jamaica, where a monotonous diet of sorghum was not imposed, it was always appreciated as food. Historian Barry Higman cites a slave hymn sung there in praise of guinea corn.[56] Only in Haiti does the cultivation of sorghum for food today remain significant.

A COLONY IN SEARCH OF A COMMODITY

From its founding in 1670 the Carolina colony was conceived and chartered as a wholly commercial enterprise. The Barbadian planters who settled it sought to enrich themselves and their English investors by building extensive plantations cropped with export commodities such as sugarcane, already a proven success in Barbados. Carolina was not planned as a political or religious haven; nor could it boast of the mineral resources that might attract fortune hunters in search of precious metals or plunder. The Carolina colony began as a plantation society, with enslaved Africans present from the outset. But its founding conceits stalled over the lack of a viable and profitable staple. Not long after their arrival, the Barbadians abandoned sugarcane because it would not prosper in the climate and soils of the Carolina lowcountry. They turned instead to the slave trade of Native Americans as a faster route to wealth.[57]

The search for a plantation commodity that could be grown for export and profit dogged the founding generation of slaveholders, who faced the far more urgent need to feed themselves. Carolina's investors initially subsidized the colonists with food imports from England, but by the middle of the 1680s the backers' willingness to extend additional support came to an end.[58] Meanwhile, arriving settlers were advised to carry at least eight months' worth of supplies. The influx of English migrants, including displaced French Huguenot tradesmen, nevertheless continued, causing the white population to swell. As the imported food rations dwindled, the settlement tilted toward starvation. In order to avert a crisis, colonists were forced to find ways to provision themselves with locally produced food.

In the midst of this situation, the plants and foods growing in plain sight on the subsistence plots of African (and Amerindian) slaves must have attracted interest. Previously ignored or disdained by whites who clung to traditional European foodways, these crops eventually formed the foundation of a new system of self-sufficiency in food. Along with Amerindian maize, beans, and pumpkins, several African plants contributed to the crop assemblage adopted by whites. For instance, a visitor to the Carolina colony in 1680–82 observed different kinds of legumes growing there and noted "many other kinds proper to the place and to us unknown." His list included two beans of African origin, the "callavance" (black-eyed pea) and "bonavist" (the lablab bean).[59] Around 1700 John Lawson mentioned several other widely consumed plants of African origin: "guinea melon" (possibly African muskmelon) and "guinea squash" (African eggplant) growing in the colony. Two decades later Mark Catesby observed that Carolina slaves planted in their food fields both sorghum and millet.[60]

By the beginning of the eighteenth century, colonists had also learned the value of African sorghum as an animal feed. John Lawson, for instance, observed that "*Guinea* Corn, which thrives well here," was grown as provender for hogs and poultry.[61] Catesby added to Lawson's observations and underscored the role of slaves in sorghum's establishment and use. While noting that maize was more commonly grown than sorghum in the Carolina colony, Catesby added that "little of the grain" of guinea corn "is propagated, and that chiefly by Negroes, who make bread of it and boil it ... its chief use is for feeding fowls, for which the smallness of the grain adapts it. It was at first introduced from Africa by the Negroes."[62]

The "almost round white" black-eyed pea described by Hans Sloane was another food plant that caught the attention of Carolinian planters. It thrived in tropical and subtropical areas and was grown for slave subsistence in Brazil, the Caribbean, and mainland North America. In the words of one historian, "everywhere African slaves arrived in substantial numbers, cowpeas followed."[63] The English initially referred to it as calavance, quite likely derived from the generic Carib Indian word for peas and beans, *calaouana*.[64] Africans, and later, slaves, cultivated the crop for its protein-rich bean and used the mineral-rich leaves as greens. The African practice of intercropping black-eyed peas with cereals was also observed in the Carolina colony. The nitrogen-fixing legume grows quickly, suppresses weeds, and improves the yields of crops grown around it. The practice of leaving cattle to graze on the plant's stems and vines or immature pods, and broader recognition of its value in the Americas as livestock fodder, is likely responsible for the plant's alternative names in English (cowpea), Portuguese *(ervilha de vaca),* and Spanish *(chícharo de vaca).*[65]

Just as the cowpea was loaded as provision on slave ships to the Americas, Southern planters raised the same pea for export to food-deficit sugar islands in the West Indies.[66] There, the crop was destined as feed "for the poorer sort of white people and for negro slaves."[67] In a survey of crops cultivated on Carolina plantations between 1730 and 1776, historian Philip Morgan found it ranked third among the crops plantations produced, after maize and rice, respectively. This export trade came to a halt with the upheavals of the Revolutionary War when the colonists looked to the cowpea as an important provision for the rebel commissariat.[68]

Sesame was also considered for its commercial potential in the colonial period. At the time when the Carolina lowcountry was being transformed into rice plantations, a colonial official named Thomas Lowndes believed that sesame oil "would make the barren pine-lands as valuable as the rice-fields."[69] In 1730 he sent samples of seed and oil from the colony to England. Sesame's usefulness as "Sallet-Oil" was already recognized in the metropole, but not yet its potential as a vegetable-oil substitute for cooking food in animal fat.[70] Lowndes was motivated by the Crown's fiscal incentives to find alternatives that reduced dependence on olive oil, which was imported from rival Catholic countries. While his schemes came to little and sesame did not develop into a commercial crop in the plantation period, it nonetheless remained significant in the foodways of the enslaved.

The interest of white Carolinians in all these foods marked the beginning of a signal shift from reliance on European foodways to the subsistence staples of Africa and America. And it was during this transition that planters began to explore the commercial potential of some of these slave foods. The search for a viable cash crop, in effect, ended in the food plots of the enslaved.

One subsistence crop emerged fairly quickly to play an important role in the economic life of the Carolina colony—rice. As early as 1648, Virginia colonists had consulted their slaves about the potential of rice as an export crop. One letter sent to England boasts that "we perceive the ground and Climate is very proper for it [rice] as our *Negroes* affirme, which in their Country is most of their food."[71] Just four years into Carolina's settlement, the colony sent "some rice which is grown on the soil to Barbados."[72] A ready market for food existed on the island and other English sugar estates in the Caribbean. But rice was not cultivated in England (or France) in the seventeenth century. In the 1670s, as in Virginia a generation earlier, the only people present in the colony familiar with its cultivation and processing were enslaved Africans who came from rice-growing societies.

Within one generation of the founding of the Carolina colony, Lawson remarked on the diverse types planted, noting that "there are several sorts of Rice, some bearded, others not, besides the red and white; But the white Rice is the best."[73] The cultivated red rice is certainly a reference to African *glaberrima*. Lawson's remarks were made at the crucial historical juncture when rice was shifting from a subsistence staple to a plantation commodity.

In the first years of the Carolina colony, rice had been introduced along with other African foodstaples. Catesby, visiting the colony in the 1720s (at a time when Carolina's rice exports had reached nearly ten million pounds a year), wrote that the first commercial attempts at its cultivation were disappointing because inferior seed was used, "it being a small unprofitable Kind little Progress was made in its Increase."[74] Catesby's comment suggests African rice, as its grains are smaller and lower yielding than the Asian species. The African species was almost certainly the initial seed source of the Carolina colony.

Rice cultivation began at a time when Carolinians had not yet determined a viable export commodity and when they depended principally on imported food subsidies. During this period of chronic scarcity, rations for slaves were undoubtedly meager, if not entirely withheld. Thus, African and Native American slaves would have had to marshal their own energy and expertise to satisfy their subsistence requirements. Africans for whom rice was previously a dietary staple were able to adapt its cultivation to the diverse and favorable growing conditions of the Carolina lowcountry.[75] Whites most likely discovered rice already growing in slave subsistence fields and, when they understood the crop's commercial potential, relied on the knowledge and skills of Africans from rice-growing societies to expland its cultivation. One schoolbook image from the 1930s captures the ambiguity that would have accompanied the transfer of knowledge from enslaved person to slaveholder and a food crop's transformation into a commodity (figure 8.1).

From its founding, enslaved Africans were present in the Carolina colony; slave imports increased the settlement's population. In the 1690s Africans formed one-fourth to one-third of newcomers; by 1710 blacks were the majority.[76] The continual arrival of slave ships landed not only newly enslaved peoples, but occasionally African foodstaples that remained uneaten at the end of the voyage. Peter Collinson described an episode that was undoubtedly repeated often in the first decades of Carolina's existence: "About this time [in the 1690s] a *Portuguese* vessel arrived, with slaves from the east, with a considerable quantity of rice, being the ship's provision: this rice the *Carolinians* gladly took in exchange for a supply of their own produce.—This unexpected cargo was distributed, which gave new spirit to the undertaking, but was not sufficient to supply the demand of all those that would have procured it to plant."[77]

The account documents the arrival of plantable seed rice discovered as surplus provision aboard a slave ship from a rice-growing area of West Africa. As

FIGURE 8.1. The initial planting of rice in Charles Towne, Carolina colony, ca. 1680.

SOURCE: Depicted in the children's book, Petersham, *Story Book of Rice.*

Collinson suggests, whites already understood that rice grew in the lowcountry swamps, and that new seed was in great demand but short supply. The unfortunate shortfall Collinson describes underscores the unique dilemma of the colony's early period. In order to promote a promising export commodity, significant reserves of seed rice would need to be accumulated and held despite the immediate pull of a subsistence crisis that would have it all distributed as food for hungry settlers.

Deliberate imports of seed rice to Carolina date to the 1690s, and by the turn of the century John Lawson could describe many of the varieties planted. The decisive turn toward rice as an export commodity was reached in this period as planters recognized the crop's suitability for commercial development. While the higher-yielding Asian species was destined to replace *glaberrima* on Carolina rice plantations, the arrival of seed rice on a slave ship suggests the way the foodstaple initially arrived in the colony along with enslaved people who knew how to plant and mill the grain. The incident moreover substantiates the claims of Maroons in South America that rice seed arrived on a slave ship.

The presence of rice and other African foodstaples in subsistence fields actually holds the key for understanding the process by which slaves instigated cultivation of African crops. The invention of the Carolina rice industry out of its quiet origins in slave food fields paralleled an explosive growth in the colony's enslaved population. By 1720, a decade after Africans had surpassed whites in number, rice had evolved into a fully sustainable export commodity. It was Carolina's chief export by the middle of the century, enriching plantation owners. Nevertheless, from its humble beginnings Carolina rice culture depended on what the first generation of enslaved rice growers already knew—that the cultivation, processing, and cooking methods informing its development were African.

Rice cultivation symbolized not only the transfer of African seed to the colony, but the simultaneous migration of an entire African agricultural and processing technology by enslaved African rice growers. The African mortar and pestle, for instance, was the only method known in colonial America for removing the grain's hulls. In the background of Carolina's formative period of rice development are slaves who used traditional African tools for pounding and winnowing plantation harvests and African techniques for irrigating and managing the colony's tidal and inland rice fields. Despite the formidable trajectory of the rice economy, and the burden the crop's transformation from subsistence staple to plantation commodity imposed on the enslaved, social memories of the original African contribution survive. The rice-growing Gullah population of the South Carolina and Georgia Sea Islands honor their African forebears by calling rice not by its English name, but by its Mande-language family name, *malo*—a name that persisted well into the twentieth century.[78]

The Gullah reverence for rice speaks to the African contribution to Carolina's culinary history. The foods that characterize the region's cuisine today are principally African and Native American. South Carolina's foodways owe their uniqueness to the fusion of African and Amerindian crops that slaves planted in their food fields. These foods quietly made their way from the plot to the plate to become the dietary and commercial staples for the entire colony.[79]

The trajectories of each of these African crops in Carolina are ineluctably tied to the institution and processes of the transatlantic trade in human beings. Slave ships carried the cultigens as rations for their enslaved captives. From food supplies occasionally remaining from slave voyages, uprooted Africans accessed the seeds and pioneered their cultivation as subsistence staples in food fields, where slaveholders discovered them and at times exploited their commercial potential. The ownership of human beings imparted to slaveholders the right to appropriate the very knowledge systems, practices, and cultivation methods that slaves used to secure their daily sustenance. Property rights gave plantation owners the power to claim that knowledge as their own and transmute it over time as proof of their intrinsic superiority and ingenuity. Slavery signifies not only an appropriation of the body and its labor, but also of the knowledge and ideas held by enslaved human beings. Significantly, it enabled the slaveholder to trade occasional favorable treatment for the knowledge and skills of the enslaved person's mind.

NINE

African Animals and Grasses in the New World Tropics

The number of sheep in the Americas eventually became immense, but they were slower to adapt than most of the other kinds of European livestock. They did not do well in the Caribbean islands or in hot, wet lowlands. . . . The areas of Brazil settled by the Portuguese in the sixteenth century were much too tropical for sheep, except in the captaincies of the central-south—Rio de Janeiro and São Paulo.

ALFRED W. CROSBY
Columbian Exchange

It was not so much the Negro's brawn as his skill that gave value to his services.

WESLEY FRANK CRAVEN
"An Introduction to the History of Bermuda"

FOOD CROPS WERE NOT THE only African biota to leave their imprint on the Americas. Packed onto ships arriving from Africa were livestock and small animals that captains purchased as fresh meat for the crews' Atlantic crossing. In an era before refrigeration, African food animals were bought live and slaughtered as needed. Their flesh provided relief from the salted meats and fish boarded in Europe. African livestock were also exported to the Americas as breeding stock. The transport of live animals required sufficient supplies of fodder and bedding to maintain them en route to the Americas, and indigenous African pasture grasses, the natural food for most of these animals, were an abundant and plentiful resource. Not surprisingly, these pasture grasses diffused to the New World along the same routes as the animals that consumed them. The grasses formed a critical component of the livestock economies that developed in tropical America.

Early Portuguese voyages of navigation along the Guinea coast frequently traded for African food animals in order to obtain fresh meat. Valentim Fernandes recorded (ca. 1506–10) that Portuguese mariners purchased live animals, small and large, from the livestock-holding societies of the Upper Guinea Coast. Captains of vessels from other European countries followed the Portuguese example. The abundant livestock herds encountered in southwest Africa prompted Edward Michelburne to write in 1604 that the "natives brought us more cattle and sheep than we could use all the time we remained there, so that we carried fresh beef and mutton to sea with us."[1]

No livestock area prompted as much commentary as Senegambia, the region where European ships journeying south along the African coast found cattle in considerable numbers. A trade in livestock developed at an early date. Englishman Richard Jobson, who visited the Gambia River in 1620–21, wrote of the large numbers of cattle tended by Fula herders. On many occasions, Jobson's crew bought "beeves [cattle], goates, [guinea] hennes, and aboundance of Bonanos... Country pease, and other graine," tasted the local rice ("even to us, it is a very good and able sustenance"), and sampled palm wine and local (sorghum) beer. Lemos Coelho, a Luso-African trader who operated in Senegambia and Guinea-Bissau between the 1640s and 1660s, remarked on the ease of purchasing food animals at low price and on the robust trade in animal hides. In an average year he reported that the English took away some fifty thousand hides from the Gambia River.[2]

In the early period of Atlantic commerce, the Cape Verde and Canary islands became important provisioning stops for ships bound for Guinea, New Spain, Brazil, and the Caribbean. In the decades before Europeans reached the Americas, Iberians inaugurated in the islands a pattern that would be repeated in the New World by populating the islands with people, plants, and animals from the African mainland. Positioned both near Africa and along crucial navigational routes, the islands served as emporia for water, salt, foodstuffs, and live animals. Not all the livestock raised on the islands originated on the Iberian Peninsula, as is commonly assumed. The African mainland also contributed animals that formed part of the herd composition on these Atlantic islands. From as early as the 1470s, raids along the African coast repeatedly netted Berber captives, cattle, small livestock, camels, and horses for the Canary Islands. Raids were relatively easy to organize: Lanzarote, the easternmost Canary Island, is located a short distance from southern Morocco. The Spanish referred to the narrow stretch of water separating the island from the African mainland as the Mar Pequeña, or Little Sea.[3]

The Cape Verde Islands offered the Portuguese similar opportunities. They are positioned only three hundred miles from Senegambia, one of Africa's notable livestock areas. In purchasing live animals in Senegambia, Cape Verdean traders such as Almada and Coelho were undoubtedly following precedent by supplementing island supplies with livestock brought from nearby sources. Proximity to the African mainland and its animals, food crops, and peoples contributed in part to the strategic importance of the archipelago to Atlantic commerce. Many vessels in transit called at the principal island, Santiago, to purchase live cattle, sheep, and goats, among other provisions, to replenish their larders before returning to sea.

As the Atlantic slave trade intensified, so did the demand for livestock as a source of fresh meat for crews. The Portuguese slave trader Manuel Bautista Pérez purchased live cattle in Cacheu (Guinea-Bissau) during his slaving expedition of 1617–18.[4] During two slave voyages to Guinea (1678–79, 1681–82), Jean Barbot's ship stopped in Senegambia, where he noted that "the crews of ships come for provisions, the necessities of life being found in abundance here."[5] "A fine cow," he added, "costs only two pieces-of-eight." The king kept "more than 5,000 animals" as did each of his noblemen. There were "herds of 200–300, with often only one black in charge of them."[6] Livestock raised on the highlands of Angola similarly supplied Luso-African merchants based in Luanda, a Portuguese administrative post and slave port.[7]

The volume of live animals purchased by a single slave ship could be substantial, as befitting the duration of a transatlantic voyage. At the Danish stronghold of Fort Christiansborg on the Gold Coast, the eighteenth-century slave ship *Fredensborg* purchased in a single transaction thirty chickens and three goats; in another, ninety-eight guinea hens, seven goats, two tortoises, seven pigs, and one cow. *The Diligent* bought live pigs, goats, sheep, and cattle at Whydah (Dahomey) prior to heading out to sea.[8]

Africans transported these animals to waiting European vessels in their canoes. In the early sixteenth century Valentim Fernandes described the canoes, carved whole from a single tree trunk, which could accommodate as many as twelve men and several cattle in one trip. In the following century Jean Barbot provided an illustration of Africans ferrying live cattle to slave ships off the coast of Sierra Leone (figure 9.1).[9] Once boarded, smaller animals such as pigs and poultry were kept in cages suspended over the sides of the ship. Larger animals were penned as deck cargo.[10]

The commerce in live animals included Africa's distinctive guinea fowl *(Numida meleagris)*. Richard Jobson mentioned seeing the "Ginney Hennes" along the Gambia River.[11] Father Jean-Baptiste Labat, drawing upon the works of Dominican missionaries and travelers to the Guinea coast, wrote in 1728 of the ways Africans tamed and managed flocks of guinea fowl: "The *Pintado,* or

FIGURE 9.1.   *Negros going aboard Ships with Provisions in their Canoos*, Gold Kingdom of Guinea, late seventeenth century.

SOURCE: Barbot, "Description of the Coasts of North and South Guinea," in Churchill, *Collection of Voyages and Travels*, vol. 5, pl. E, p. 99.

Guinea-Hen, is seen through all this Country in great Numbers. They are naturally wild, but easily tamed, and are often brought over to *Europe*. In shape they resemble a Partridge, but are larger. Their Feathers are of a dark Ash Colour, so regularly spotted with White, that it renders their Plumage very beautiful.... Their Flesh is white and good. . . . They keep them in Flocks of two to three hundred together, and the Negros run them down with Dogs. If taken young, they grow tame as poultry."[12] Small and portable, the bird was known to "thrive well aboard of ships, and live long."[13]

The earliest confirmed evidence for the introduction of the guinea fowl to the New World is from Hispaniola, where it arrived in 1549 aboard a vessel that had left the Cape Verde Islands. A Jesuit priest claimed that the guinea fowl arrived in the Caribbean even earlier, on ships that carried the first boatloads of African slaves.[14]

The guinea fowl was also present in Brazil at an early date. Georg Marcgraf, a member of the first scientific expedition to Dutch Brazil (1638–44), mentioned it among the faunal species he found there and provided the first illustration of the guinea fowl in the Americas (figure 9.2). He acknowledged its African provenance, calling it the "African chicken."[15] The common names by

FIGURE 9.2. Illustration of the African helmeted guinea fowl, by Georg Marcgraf, ca. 1640.

SOURCE: Marcgrave, *História natural do Brasil*, 192.

which the guinea fowl is still known in Brazil—*galinha* (chicken) *d'Angola* and *galinha de Guinea*—incarnate a social memory of the species' introduction from Africa.

The guinea fowl formed a significant component of the small animal stock that plantation slaves kept and raised in the small plots surrounding their dwellings. They occasionally sold them to their masters. The guinea fowl was also featured in dishes that enslaved cooks prepared for the Thomas Jefferson family at Monticello. Today it remains an important poultry species reared by African Americans in the southern United States. It is also kept in many other parts of tropical America as a food animal. In Brazil, the guinea fowl is used in

the *Rice Plant*

a *Sestro Sheep*

FIGURE 9.3. Illustration of sheep and African rice, Sierra Leone, late seventeenth century.

SOURCE: Barbot, "Description of the Coasts of North and South Guinea," in Churchill, *Collection of Voyages and Travels*, vol. 5, pl. F, opposite p. 128.

the liturgical practices of Africa-based religions such as candomblé. As a food animal and ritual symbol, the guinea fowl represents both an important African faunal contribution to tropical America and provides diaspora descendants a direct living link to their African heritage.[16]

European ships transported other African food animals to plantation economies. Of three distinct breeds of sheep introduced to Antigua, Barbuda, and Anguilla, two came from West Africa during the early colonial period.[17] One of these was the African hair sheep, whose distinguishing characteristic Lemos Coelho described in seventeenth-century Senegambia: "There are many sheep, which are unlike those of Europe in that they have no wool, but their meat is most tasty."[18] Barbot featured the same woolless sheep alongside African rice in a drawing of Sierra Leone (figure 9.3). The hair sheep was present at an early date as a food animal in tropical America. Writing about Jamaica in the 1680s, naturalist Hans Sloane claimed that the island's sheep were of a breed that was

FIGURE 9.4.   Illustration of African sheep, Dutch Brazil, by Georg Marcgraf, ca. 1640.

SOURCE: Marcgrave, *História natural do Brasil,* 234.

originally African. Another account of the island a century later notes that the sheep "have short hair instead of wool.... They are small, but very sweet meat." Raised as a food animal, and well acclimatized to the tropical environment, by the early seventeenth century the African sheep were found in the West Indies, Guiana, and Brazil.[19]

African sheep were also present in Barbados in the 1640s, introduced—as Richard Ligon noted—by ships departing Upper and Lower Guinea: "Other sheep we have there which are brought from *Guinny* and *Binny,* and those have hair growing on them, instead of wool; and liker goats than sheep, yet their flesh is tasted more like mutton than the other."[20] At the same time, Georg Marcgraf in Dutch Brazil depicted the woolless African sheep among the fauna new to him (figure 9.4). The Portuguese common names for the hair sheep—*carneiro de Guiné* and *carneiro d'Angola*—similarly recall its African origin. The hairy coat deprived the sheep of any economic value as a wool producer, but it nonetheless served as a suitable meat animal for the New World tropics, as Griffith Hughes recognized in the mid-eighteenth century: "The sheep that are natural to this climate ... are hairy like goats. To be covered with wool, would be as prejudicial to them in these hot climates, as it is useful in winter countries for shelter and warmth." Today the African hair sheep is still raised in Barbados, some parts of the Caribbean, and along the north coast of South America.[21]

Marcgraf drew yet another animal species—the *porco de Guiné*, a short-haired and reddish-colored domesticated pig that had been introduced to northeastern Brazil. Marcgraf claimed that the pig came to Brazil from Guinea.[22] As with the guinea fowl and hair sheep, this swine breed was established in the early plantation period for its suitability as a food animal in the humid tropical climate. It provided yet another source of meat on a continent where Native Americans had domesticated few animals. When Europeans and Africans arrived in the Americas, the only domesticated Amerindian animal species were the llama, alpaca, dog, guinea pig, turkey, and Muscovy duck; wild game supplied most of the Amerindian need for meat and hides.[23]

THE ATLANTIC ISLANDS AND LIVE ANIMAL INTRODUCTIONS

The strategic location of the Atlantic islands facilitated their emergence not only as provisioners to passing ships, but also as suppliers of breeding stock for the New World tropics. Reliance in part on African animals to populate island herds makes it likely that African cattle also arrived in the Americas. Portuguese and Spanish sources underscore the significance of the Atlantic islands for livestock introductions to the New World.[24] For example, cattle from the Cape Verde Islands were among the founding stocks carried to sixteenth-century Bahia, Brazil.[25] The ship that brought Richard Ligon to Barbados in 1647 stopped in Santiago to purchase fifty head of cattle and eight horses, which were sold at the end of the three-week Atlantic crossing.[26]

Cattle were in great demand in the New World. On sugar plantations, they were imported for turning millstones that crushed cane, for transport, and to produce manure critical for maintaining soil fertility (figure 9.5). Besides meat and dairy, cattle provided many other useful products. Their hides gave a versatile fabric, equivalent in its diverse applications to the uses of plastic in our own time. The animal's fat yielded tallow for making candles and soap; rendered, it provided fat for cooking.

European cattle, adapted to temperate climates and temperatures, did not thrive in the humid tropical lowlands. As Alfred Crosby points out, "In steamy Brazil and the Colombian and Venezuelan llanos, Iberian cattle took generations to adapt; but in the higher country they exploded in numbers."[27] African cattle may have well made decisive genetic contributions to the breeds that proved suitable to the climate of Barbados and other New World regions. Eu-

FIGURE 9.5.   Cattle use in a Brazilian sugar mill depicted in *Brasilise Suyker-werken*, by Simon de Vries, 1682.

SOURCE: Courtesy of the John Carter Brown Library at Brown University, Providence, Rhode Island.

ropean ship captains purchased cattle in Guinea, as they did other food animals. From an early date African cattle contributed to the herds raised on the Atlantic islands for sale as breeding stock to passing ships. Senegambia's hardy *n'dama* cattle likely figured in the development of a breed suitable to the climate of Barbados and other lowland tropical regions.

Other domesticated animals from Africa appeared early in the development of New World plantations. Besides the guinea fowl and woolless sheep, there is an unexpected but singular addition: the one-humped dromedary camel. The camel was already a feature of the Barbadian landscape when Ligon arrived in 1647:

> If I shall begin with the largest [animals], first I must name Camels, and these are very useful beasts, but very few will live upon the Island: divers have had them brought over, but few know how to dyet them. Captain *Higginbotham* had four or five, which were of excellent use, not only for carrying down sugar to the bridge, but of bringing from thence hogsheads of Wine, Beer, or Vinegar, which horses cannot do, nor can Carts pass for

FIGURE 9.6.   Richard Ligon's map of Barbados, ca. 1647.
SOURCE: Ligon, *True and Exact History,* frontispiece.

Gullies, and *Negroes* cannot carry it, for the reasons afore-mentioned; a
good Camel will carry 1600 l. weight, and go the surest of any beast.[28]

Significant here is Ligon's admission that "few know how to dyet them." The
comment reveals planter unfamiliarity with camels. Some of the slaves who were
brought to the island, however, did have knowledge and experience raising this
notoriously hard-to-handle animal. The map Ligon attached to his account
shows camels—one clearly tended by an African—two types of swine, donkeys,
and cattle (figure 9.6).[29] Barbados was not the only site of the camel's New World
introduction. They were present in Saint Domingue in the 1750s.[30] Two centuries earlier, camels had been brought to Peru as pack animals. Introduced in
the 1540s shortly after the Spanish conquest, they were used to transport heavy
loads across the coastal desert. The experiment was evidently not a lasting success: the camel's presence in Peru ended about 1615, when runaway slaves reportedly killed the last remaining specimen for food.[31]

The camels that were imported to sixteenth- and seventeenth-century trop-

FIGURE 9.7.   Camel transporting bananas with donkey in background, Canary Islands, n.d.

SOURCE: Simmonds, *Bananas,* pl. 48, opposite p. 272.

ical America did not arrive on ships from Arabia, but came rather from the Canary Islands. In the millennium prior to the expansion of Islam, camels had been repeatedly introduced to Africa from the Arabian Peninsula. In the Greco-Roman period, the camel linked North Africa to the inland Niger Delta through the caravan routes that passed through desert outposts controlled by the Garamantes and the Fur people of Sudan's Darfur region. Camels also became an important component of livestock herds between southern Morocco and the Senegal River. They are evident in the archaeological record of the Senegal River valley by 250–400 C.E.[32]

In the fifteenth century, the camel was brought across the narrow stretch of water that separates southern Morocco from the easternmost Canary Islands. Camels arrived with the first Spanish slave-raiding ventures to the African mainland, which captured both Berbers and their domesticated animals.[33] The camel remains to this day a curious feature of Fuerteventura's landscape (figure 9.7).[34]

The importance of livestock adapted to the climate of tropical lowland America is suggested through early commentaries on tropical pasture grasses that we now know to be of African origin. Summarizing the historical record, geographer James Parsons considered the diffusion of African grasses to the New World to be one of the "most rapid and significant ecologic invasions in the earth's history." He called the botanical dispersal the *"Africanization* of New World tropical grasslands" without engaging the possible role of Africa's animals and enslaved herders in adapting them to a new continent.[35] The early presence of the nutritive African forage species lends supporting evidence for the likelihood that African livestock were present in the early herd composition of tropical America, while showing yet another facet of Africa's broader botanical legacy.

Indigenous American grasses proved less significant than African grasses to livestock development in the lowland tropics. Only within the temperate-zone Great Plains of North America or on the Argentine pampas do we in fact find productive forage species with high grazing value. But in the lowland tropics, the "African grasses stand up better to grazing and have higher nutritive values than native American species."[36] This is in part due to the fact that the African pasture grasses coevolved over millennia with livestock raised and bred by herding peoples. The New World tropical lowlands did not have large domesticated animals or the nutrient-rich pasture grasses necessary for raising them. Instead, the more vigorous African species served as the foundation for livestock development in these areas.

With their seeds and cuttings readily disseminated by animals transported to the New World, the African grasses took root and flourished. Five species in particular transformed the grazing economies of New World tropical and subtropical regions: guinea grass *(Panicum maximum),* Pará or Angola grass *(Brachiaria mutica),* molasses grass *(Melinis minutiflora),* Bermuda grass *(Cynodon dactylon),* and jaragua grass *(Hyparrhenia rufa).* As the common names Pará and Bermuda grass indicate, some of the African species were present in the Americas from such an early date that they were thought to be natives; this belief persisted into the twentieth century.

Guinea grass *(Panicum maximum)* was among the most important pasture grasses in plantation economies. It grew in well-drained fertile soils and reached a height of six feet or more, which made it a useful cut green fodder for stabled animals, especially dairy cattle, horses, and mules. Guinea grass probably reached Brazil's Northeast by the 1600s. The grass was sustaining cattle on Barbadian sugar plantations in the 1680s and was likely the pasture grass that Hans Sloane collected in Jamaica during that decade.[37] In the eighteenth century Bryan Edwards—planter, politician, and proslavery advocate who lived on Jamaica between 1759 and the 1770s—credited a ship captain, who purportedly intended its use as birdseed, with the plant's introduction half a century earlier:

> Perhaps the settlement of most of the north-side parishes is wholly owing to the introduction of this excellent grass, which happened by accident about fifty years ago; the seeds having been brought from the coast of Guiney, as food for some birds which were presented to Mr. Ellis, chief-justice of the island. Fortunately the birds did not live to consume the whole stock, and the remainder, being carelessly thrown into a fence, grew and flourished. It was not long before the eagerness displayed by the cattle to reach the grass, attracted Mr. Ellis's notice, and induced him to collect and propagate the seed; which now thrive.[38]

Edwards's account is of interest for several reasons. He notes the plant's casual introduction, but emphasizes the role of a ship captain and colonial official in the dispersal and propagation of the seed. Edwards dated the arrival of guinea grass to the early eighteenth century. The subsequent economic impact of guinea grass in Jamaica had made the manner of its arrival worthy of Edwards's attention. He ranked "Guiney-grass . . . next to sugar-cane, in point of importance, as most of the grazing and breeding farms, or pens, throughout the island were originally created and are still supported, chiefly by means of this invaluable herbage."[39] At the time Edwards wrote, guinea grass was planted as a quality forage crop in many parts of tropical America. The plant's arrival via a ship captain who proffered it as birdseed does not offer a convincing explanation for the crop's multiple New World introductions and broader biogeographical distribution. However, Edwards inadvertently touched on a more credible scenario when he repeated Hans Sloane's seventeenth-century observation that "the sheep of Jamaica, according to Sloane, are from a breed originally African."[40] The frequent export of African livestock to many different New World plantation societies makes it far more plausible that guinea grass traveled to tropical America on animal hoofs and in the bedding and feed that accompanied them on Atlantic crossings.

Guinea grass was prized as forage and fodder throughout the Caribbean and mainland tropical and subtropical America. It was planted in Portuguese Brazil, the Spanish colonies of northern South America, French Saint Domingue, and the English colony of Carolina. Henry Laurens, the eighteenth-century South

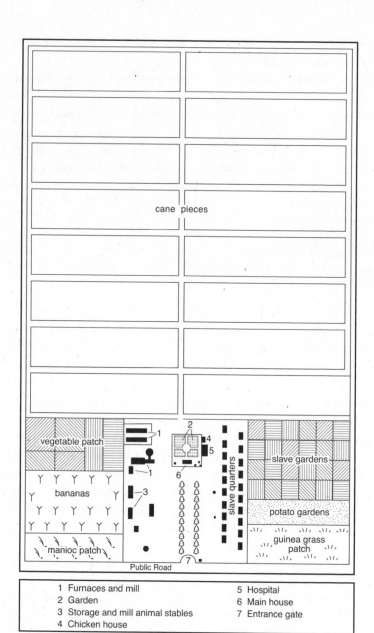

cane pieces

vegetable patch

1

2

4

5

slave gardens

bananas

3

6

slave quarters

potato gardens

manioc patch

guinea grass patch

7

Public Road

1 Furnaces and mill	5 Hospital
2 Garden	6 Main house
3 Storage and mill animal stables	7 Entrance gate
4 Chicken house	

FIGURE 9.8. Guinea grass field depicted in idealized sugar plantation, Saint Domingue, 1796.

SOURCE: Watts, *West Indies,* 390.

Carolina planter and merchant, claimed to have imported the seed "from remote parts of the globe."[41] On sugar islands—where arable land was scarce but the need to provision cattle, oxen, and horses critical—plantations frequently reserved a patch of land for the cultivation of guinea grass as fodder. It is featured in one sketch of plantation land use from prerevolutionary Haiti (figure 9.8). Pastures were sown to guinea grass in the Leeward Islands and other semiarid parts of the Caribbean.[42]

Sloane collected guinea grass (known in the 1680s as Scotch grass after the Barbadian district where he first encountered it) in Barbados and Jamaica. He described the grass as a cultivated fodder "planted in moist ground all over the island for provision for cattle." Sloane claimed this pasture grass was deliberately introduced to Jamaica: "after its being found very useful in Barbadoes, and had been there planted for some time, it was brought hither [to Jamaica] and is now all over the island in the moister land by river sides, planted after the manner of sugar canes."[43]

The second important African livestock grass was known in Brazil as *capim de Angola*—Angola or Pará grass *(Brachiaria mutica)*. Pará grass was cultivated on many Caribbean islands and in the lowlands of Venezuela, Suriname, Colombia, and northeastern Brazil. While guinea grass required well-drained soils for cultivation, Pará grass prospered on the moist poorly drained bottomlands of tropical America. It was grown in pastures as forage for cattle or was cut as a fodder for stabled animals.[44]

The German naturalists Johann Baptist von Spix and Carl Friedrich Philipp von Martius observed the cultivation of many different forage grasses in their early nineteenth-century visit to Brazil—species that they collectively referred to as guinea grass. These fodder species were planted on the outskirts of cities, within reach of the urban markets that demanded feed for their transport animals and dairy cattle. "Fresh grass *(capim),* which is grown in gardens in the neighbourhood, is brought to market as fodder for the cattle, especially for horses and mules."[45] Slaves cut and bound the grass into long bundles and often transported them to markets on their heads. This feature of urban life in Brazil was frequently depicted by visiting painters. One was Carlos Julião, an Italian artist who resided in the Portuguese colony during the last quarter of the eighteenth century (plate 6). French painter Jean Baptiste Debret, based in Rio de Janeiro between 1816 and 1831, was more explicit about the grass involved in the fodder trade. The commentary that accompanies his painting (figure 9.9) identifies the fodder as Angola grass *(capim de Angola).* He wrote of the enormous demand for *capim* by city dwellers on behalf of their animals. It was bundled in sheaves and brought to town on the heads of slaves, on the backs of

FIGURE 9.9.   *Vendedores de capim e leite* (Vendors of [Angola] grass and milk),
Brazil, by Jean Baptiste Debret, ca. 1821–25.

SOURCE: Debret, *Viagem pitoresca e histórica ao Brasil,* vol. 1, pl. 21, following p. 164.

donkeys, and in carts. Debret's image of Angola grass presents in a compressed
pictorial space the fundamental relationship between this African grass, the livestock economy, dairy products, and slaves.[46]

Pará or Angola grass may also be the cut green fodder illustrated in a mid-
nineteenth-century painting from Cuba. In Cuba the fodder cut from maize
leaves and other suitable grasses is known generically as *maloja* and those who
market it as *malojeros.*[47] Plate 7 depicts vendors selling a long cut grass for fodder in an urban setting. The grass is not maize stalk; more likely it is Pará or
guinea grass. Both were well established pasture grasses on the island.[48]

Bermuda grass was named after the arid Caribbean island, which was uninhabited when the Spanish let loose small food animals in the early sixteenth
century to proliferate as a ready meat supply for passing ships. The English introduced cattle there the following century via ships that left the Canary Islands. Long believed to be a native species, Bermuda grass is actually African.
The presence of the species on the island that gave the grass its name suggests
that African food animals, acting as seed vectors, were among the livestock deposited on Bermuda at an early date. Bermuda was a frequent stopover for English colonists bound for Virginia during its initial settlement period. Both

Bermuda and Virginia in turn contributed cattle to the Carolina colony.[49] By the middle of the eighteenth century, Bermuda grass had spread throughout the lower South, possibly along with cattle and hogs introduced from the island or directly from Africa.[50]

A few African crops that served the dual purpose of human and animal feed provide a glimpse of the intermediary role slaves likely played in these plants' subsequent trajectory in plantation societies. The African cowpea or black-eyed pea became one of the most widely employed forage crops of the lower South. Many plantations grew the crop for forage in cattle pastures or cut its vines for hay. In fact, the English vernacular name "cowpea" underscores its important association with cattle. The cowpea remains an important rotation crop in the South, where it is known as pea hay. In the Caribbean, slaves and their descendants planted it to define the boundary of their food plots and as feed for the small animals they raised.[51]

Guinea corn was another multipurpose African crop that slaves cultivated for animal feed. Following his visit to the Carolinas and Florida, naturalist Mark Catesby observed in 1754 that "little of the grain is propagated, and that chiefly by Negroes, who make bread of it and boil it ... its chief use is for feeding fowls, for which the smallness of the grain adapts it."[52] Sorghum is today a leading source of cattle feed in many parts of the Americas.

AFRICAN AGROPASTORAL PRACTICES IN TROPICAL AND SUBTROPICAL AMERICA

European visitors to the New World were repeatedly struck by the spectacle of massive cattle herds that had multiplied from stock introduced only a short time before. Old World domesticated animals that were unknown in the Americas before Columbus—cattle, sheep, goats, horses, and pigs—ultimately came to thrive in their new environments. The huge cattle herds that were raised in the New World elicited admiring commentary. Within decades of the founding of the Carolina colony in 1670, John Lawson remarked that cattle numbers were "incredible, being from one to two thousand Head in one Man's possession."[53] In 1682, a visitor to Carolina noted the astonishing growth in cattle herds: "Not more than six or seven years past the country was almost destitute of *Cows* ... and now they have many thousand Head."[54] Carolina's herds descended not from feral cattle from Spanish Florida but from stock imported from Bermuda, Virginia, and New York. In the eighteenth century, Carolina's cattle surpluses allowed the colony to become an important supplier of salt beef to the British West Indies during the same decades that its plantations were turning to rice.[55]

Lawson's comments on Carolina cattle were echoed by other observers in the Caribbean and mainland tropical lowlands. In mid-sixteenth century Hispaniola, cattle numbers exceeded four hundred thousand head, and hides became the colony's second export.[56] Indeed, hides formed an important component of the commerce between the Spanish colonies and the metropole. By the early 1580s, Hispaniola reportedly was able to raise and slaughter enough animals to export some fifty thousand hides a year.[57] In the 1670s, Martinique supported more than five thousand cattle and twelve hundred horses. At the end of the seventeenth century there were over 130 ranches established in the Brazilian northeastern states of Piauí and Maranhão. On the eve of the Haitian Revolution, the French colony of Saint Domingue counted more than one-quarter million cattle, sheep, and goats and an additional fifty thousand horses. In the early 1800s, Alexander von Humboldt estimated the cattle population he saw on the Venezuelan savannas at over sixty thousand head.[58]

The shift to animal husbandry represented a remarkable ecological transformation, one that was abetted by the introduced African pasture grasses. The grazing lands of tropical America increasingly resembled those of Senegambia, which European navigators had long admired for the great cattle herds they supported. But the faunal and botanical contributions of the African continent were not the only ingredients informing livestock development in tropical America. The remaining components were human and cultural—the presence of enslaved livestock keepers from societies who were versed in raising cattle.[59] This point was indirectly acknowledged in the comments of seventeenth-century trader Lemos Coelho, who noted a consistent demand for slaves from Senegambia possessing such abilities.[60] It was substantiated by Père Jean-Baptiste Labat in Martinique a few decades later, when he recommended that sugar plantations seek cattle minders and horse grooms among slaves brought from Senegambia since they were accustomed to cattle herding.[61] As historian John Thornton observes, "It is perhaps no wonder that all the *vaqueros* and *ganaderos* (cowboys) on the mid-sixteenth-century Hispaniola estates were not only Africans but from the Wolof, Fula, and Mandinga areas [of Senegambia], where there was a strong equestrian and cattle-raising tradition."[62] African slaves who were livestock handlers were indeed among the first cowboys of the Americas, as some scholars have suggested.[63] It was this African faunal, botanical, and cultural legacy that contributed to livestock development in tropical America.

Most of the research on New World livestock development emphasizes the role of Iberian animals and the Peninsular ranching tradition.[64] For the most part, the scholarship fails to consider that Iberian practices principally succeeded in regions with similar temperate or Mediterranean climates to which the animals were adapted. While the European grasses that altered the landscape of California must credit the Iberian livestock tradition, there is less attention paid to the role of African species and the expertise of Africans in revolutionizing New World tropical and subtropical ecosystems. African accomplishments in cattle breeding and the sophisticated knowledge systems that informed herder practices likely contributed to the growth and vigor of New World herds. These accomplishments are not widely known but are possibly quite pertinent for understanding how Old World cattle so quickly adjusted to the New World tropics during the early period of plantation development.[65]

Slave traders operating in Senegambia evinced a keen interest in the cattle herding practices of the Fula.[66] In the early seventeenth century, Richard Jobson wrote of the "Fulbies" whose "profession is keeping of Cattle."[67] He added: "These mens labour and toyle is continuall, for in the day time, they watch and keepe them [cattle] together, from straying ... and in the night time, they bring them home about their howses."[68] Later in the century, Jean Barbot's attention was drawn to the way the Fula safeguarded their animals after dark: "at night these Africans shut them [cattle] up in pens as we do with sheep in Europe."[69]

Francis Moore, an English slave trader who lived along the Gambia River in the 1730s, elaborated on Fula livestock management. Moore discussed and illustrated how cattle were penned at night in an area corralled by thick manmade barriers of brush (see figure 3.6). He described nocturnal tethering areas where cattle were individually tied to stakes in the ground: "They have a Place near each Town for the Cattle. . . . They drive great Numbers of Stakes in Rings . . . and every Night they duly bring up the Cattle, who are so tame and well accustomed to it, that they come up with Ease; each Beast is tied separate to a stake.... After the Cattle are tied, they milk the Cows.... The Calves they wean from the Cows and keep in a Common Penn, which is made with so strong and high a Fence around it, that no wild Beast can pass it. . . . In the morning they again milk the Cows, and let them go into the Savannahs."[70] A century earlier, Jobson had written how the English visited local tethering places to buy beef or "beeves."[71]

In Senegambia these practices have remained largely unchanged over the centuries (see figures 1.2 and 1.3). The tethering of an animal to a stake has advantages beyond protecting livestock from predators: it concentrates manure, improves soil fertility, and importantly, prevents animals from wandering into planted areas. The practice is especially adapted to mosaics of land use where cultivated fields are interspersed with those in fallow. Tethering embellishes an agropastoral system in which agricultural fields rotate annually from cultivation to animal pasture. Once the crop is harvested, the field reverts to pasture; animals are moved in to clear the stubble and fertilize the soil with their manure, thus regenerating the field in advance of the next planting cycle. Moore's sketch reveals the close relationship of animal husbandry to farming in Fula villages.

Moore also observed that the cattle-keeping Fula were better farmers than the Mandinka—nonpastoralists who depended in part on the Fula to make up occasional food shortages: "Were it not for these *Pholeys* [Fula], I believe, many of the *Mandingoes* would want Sustenance."[72] Moore's comment silently acknowledges the superior crop yields made possible by the Fula system of agropastoral rotation.

The African practice of livestock tethering was reproduced in many plantation societies of the English Caribbean. There is little doubt that Africans familiar with tethering had been brought to Barbados from an early date. Lemos Coelho noted in the mid-seventeenth century that English traders operating in Senegambia—a region populated by Fula agropastoralists—"buy many blacks whom they ship to Barbados."[73] In 1830 tethering was described as a customary method to restore fertility to fallow Barbadian cane fields: "The practice is to tether cattle to stakes driven into the ground. The spot is covered with good mould, and then well littered with dry and green vegetable matter, which, with the animal manure from the cattle, makes a compost heap sufficient for a certain space of ground. When this is completed the stakes are withdrawn, and placed in another part of the field, in which the same process is renewed."[74]

This practice continued on the small subsistence plots that freed persons managed in the postemancipation Caribbean. An English agronomist writing about Barbados in 1905 reported that nearly every peasant proprietor on the island held sheep, "usually tethered to a peg while pasturing in the day time and placed under cover at night."[75] The practice offered subsistence farmers an ideal solution for maintaining the fertility of individual plots in areas where a patchwork of small farms (typically less than one acre) characterized the holdings of manumitted slaves and their descendants.

The tethering of animals forms an important component of the Antillean subsistence complex characterized by Riva Berleant-Schiller and Lydia Pulsipher. In the West Indies it is still customary on small unfenced plots to tie livestock to a peg in order to concentrate manure and to prevent animals from wandering onto the cultivated parcels of neighbors. In their broad comparison of Caribbean smallholder agropastoral strategies with those practiced on mainland tropical America, Berleant-Schiller and Pulsipher see tethering as distinctly West Indian: "The integration of livestock is characteristic of gardens everywhere in Latin America, but outside the Antilles the pens are rotated through the plot, and dung is hand carried [to the area where the fertilizer is needed].

In the Antilles the tethering on fallow supplements these techniques."[76] The system of erecting a temporary cowpen and moving it across an uncultivated field is thought to have resulted from the cultural interchange between Europeans and Africans in the New World, but pegging individual animals to a stake on small food plots is an African practice.[77]

African contributions to Caribbean agropastoral systems are also still evident in Barbuda. In the seventeenth century, the semiarid island developed into a specialized livestock economy, which supplied beef to the sugar plantations of its neighbor, Antigua. Barbuda effectively was transformed into pastureland that slaves managed. Left to provide for themselves, Barbuda's slaves adapted their cultivation practices to the availability of animal fertilizer by locating food plots in areas that were amply manured. They grew Amerindian and African foodstaples that were suited to the island's environmental conditions. Chief among these were sorghum, plantains and bananas, pigeon peas, tannia *(Xanthosoma saggitifolium)*, sweet potatoes, and diverse types of beans. Barbuda's African descendants continue to practice a variation of this agropastoral system on land that is held in common.[78]

The rotation of land between pasture and agricultural field was instrumental to the development of the Carolina colony. As in other ranching areas of mainland America, enslaved Africans in Carolina tended large numbers of cattle. But what distinguishes the colony is the reconstitution of a key feature of the Senegambian agropastoral system: the temporal shift of land use from cattle pasture to an agricultural field—in this case, practiced on soils planted to rice.[79]

In 1700 there were an estimated three thousand enslaved Africans in Carolina—comprising nearly half the colony's immigrant population. In these early years, most of the colony's slaves arrived via the British West Indies.[80] Among the Africans brought to the colony were slaves from Senegambia, Sierra Leone, and the rice-growing and herding interior. They were "imported from West Africa by the Royal African Company. In the years between 1673 and 1689, for example, the Company shipped a total of 90,000 slaves to the British West Indies. Over a tenth of these slaves came from Senegambia."[81] This observation was made at a time when the Royal African Company held a monopoly in the English slave trade. The dates span the years following Lemos Coelho's written account, which records the English demand for slaves from Senegambia—a region then known both for animal husbandry and rice production.[82] Cattle herders and rice growers may not have been a majority of the enslaved Africans present in South Carolina's formative period, but the transfer of these crucial forms of knowledge needed only the critical presence of a few skilled practitioners.

Historian Wesley Frank Craven alluded to a similar conclusion more than seventy years ago when he described the early settlement history of Bermuda. The arrival of English colonists in 1615 preceded by one year the landing of the first African slaves. In fact, Bermuda was the first English colony to receive African slaves, antedating even the notable disembarkation of Africans at Jamestown in 1617. The English settlers of Bermuda so valued the agricultural skills and experience of specific Africans that one was put "in charge of those experiments with tropical and semi-tropical plants on which were pinned so many of the adventurers' hopes." Another enslaved man, mentioned in 1618 as landowner Robert Rich's most prized possession, "was a negro especially skilled in the 'planting of West Indy plants.'" The same year documents the arrival of female slaves and the turn to a workforce increasingly dependent on slavery.

Craven recognized the abilities of individual slaves in the settlement history of Bermuda when he wrote that "it was not so much the Negro's brawn as his skill that gave value to his services." However, Craven never doubted that this knowledge had been thoroughly instilled by Europeans: "Picked up in West Indian islands, or seized with Spanish or Portuguese ships," these Africans "brought to the aid of the English colonists something of the experience and skill of their principal rivals. In return they enjoyed a position considerably better than that of later slaves."[83] But recent research increasingly suggests that enslaved Africans arrived in the New World already possessing many of the skills and knowledge systems that came to be valued by their eventual owners. The contributions of Bermuda's first slaves could broadly apply to the rest of tropical America and the formative period of many plantation societies. For among the millions of Africans forcibly dispersed to these regions were members of specialist ethnic groups and people skilled as herders, farmers, blacksmiths, dyers and spinners of textiles, and diverse craftpersons. The complex of knowledge they held was neither trifling nor incidental. It spanned vast categories of human endeavor, having produced in many of these vocations matchless levels of expertise. In the New World, Africans and African knowledge made indelible contributions to the shaping of landscapes where they lived and labored.

PLATES

PLATE 1.   Women gathering grain, prehistoric fresco from Tassili n'Ajjer, Algeria, ca. 2000 B.C.E.

SOURCE: Henry Lhote Collection, Musée de l'Homme, Paris. Photo © 1966 Erich Lessing, Art Resource, New York.

PLATE 2. *Transport des nègres dans les colonies* (Transport of negroes in the colonies), color lithograph by Prétextat Oursel, early 1800s.

SOURCE: Photo by Michel Dupuis, Ville de Saint-Malo. Reprinted with the permission of the Musée d'histoire, Saint-Malo, France.

PLATE 3. Illustration of the Danish slave ship *Fredensborg,* ca. 1785.

SOURCE: Reprinted with permission of the Danish Maritime Museum, Kronborg, Denmark.

PLATE 4. Illustration of slaves washing for diamonds at Mandango on the River Jequitinhonha in Serra do Frio, Minas Gerais, Brazil, by Carlos Julião, ca. 1776.

SOURCE: Reprinted in Cunha, *Riscos iluminados de figurinhos de brancos e negros,* pl. 42.

Figura y tamaño del Plátano Guíneo.

Figura en lo íneríor del Plátano Guineo, el qual es muy sano.

Figuras en lo exteríor, éínteríor de la Guayaba, la qual es Caliente.

PLATE 5. Illustration of *plátano guineo* (banana) and guavas, by Joaquín Antonio de Basarás y Garaygorta, 1763.

SOURCE: Basarás y Garaygorta, "Origen, costumbres, y estado presente de mexicanos y philipinos" (unpublished ms., 1763). Courtesy of the Hispanic Society of America, New York.

PLATE 6. Illustrations of grass and milk vendors, Rio de Janeiro, by Carlos Julião, ca. 1776. Note the punishment collar worn by the man on the right.

SOURCE: Cunha, *Riscos iluminados de figurinhos de brancos e negros,* pl. 34.

PLATE 7. *El panadero y el malojero* (The bread seller and the fodder seller),
Havana, Cuba, by Pierre Toussaint Frédéric Miahle, ca. 1847–48.

SOURCE: Pierre Toussaint Frédéric Miahle, *Album pintoresco de la isla de Cuba* (Havana, 1850), in
Handler and Tuite, "Atlantic Slave Trade and Slave Life in the Americas: A Visual Record," database at
University of Virginia Library, image ref.: album-13.

PLATE 8.   Illustration of female food vendors, colonial Brazil, by Carlos Julião, ca. 1776.

SOURCE: Cunha, *Riscos iluminados de figurinhos de brancos e negros,* pl. 33.

TEN

Memory Dishes of the African Diaspora

Specific foods, no matter how humble in our eyes, excite the same symbolically mediated complexity that foie gras and caviar excite in gourmands, because we are the only animals that have culture, and that's the big secret, which people are taught but do not really believe. . . . The single most important truth about human beings is the existence of culture.

SIDNEY W. MINTZ

AFRICAN INGREDIENTS AND COOKING PRACTICES gave the foodways of former plantation societies their distinctive culinary signatures. Their metamorphosis to the diaspora cuisines of today was originally mediated by enslaved women who guided modest foods out of the subsistence plot and into the cooking pot. These foodways began, to borrow the words of historian James McWilliams, as "cuisines of survival."[1] At the hearths of their dwellings and in the kitchens of plantation gentry, African women and their descendants created the fusion cuisines and memory dishes that attest to the African presence in the Americas.

A signature ingredient of the foodways of Africa and the diaspora is greens. Perhaps no other cooking traditions feature them so prominently. In West Africa alone there are more than one hundred fifty indigenous species of edible greens. Twentieth-century botanists identified more than thirty different cultivated species in the region.[2] Greens are mentioned in seventeenth- and eighteenth-century accounts of African meals. It may well be that the unidentified plant "the size of Parsley" that Pieter de Marees included in his depiction of foods grown along the Gold Coast in 1602 was one of these edible cultivated greens (see figure 3.5).[3] Africans traditionally gather many types of wild plants for food, but women are the principal experts in cultivating, cooking, and marketing leafy vegetables.[4] Greens contribute in fundamental ways to stews and the sauces that accompany starchy staples. They are commonly grown in kitchen gardens; such proximity to the hearth testifies to their dietary importance. Greens are a frequent addition to the one-pot stews that have long distinguished African cooking.[5]

Greens are served in a variety of ways: uncooked as a salad, boiled in side dishes as spinach, mixed in soups and stews or with other vegetables as potherbs, or used as garnishes. Greens generally impart a bitter taste to food, a trait much emphasized in Africa-based cuisines.[6] Some greens are prized as thickening agents for soups and stews, such as the leaves of sesame, hibiscus, jute mallow *(Corchorus olitorius),* the baobab tree, and most famously, okra.[7] Each adds a mucilaginous texture, binding the ingredients of soups and stews together. In all of these various culinary guises, greens contribute crucial stores of vitamins, minerals, and micronutrients to the diet.[8]

The centrality of greens to the food culture of sub-Saharan Africa cannot be understated. When New World manioc was introduced to Angola during the transatlantic slave trade, women experimented with the plant's leaves for food despite awareness of the poisonous root.[9] They discovered that young manioc leaves were not only safe to consume as boiled greens but were also nutritious, as one expert on the usage of manioc in Africa verifies: "The manioc root, although rich in starch, contains very little protein or fat and very small amounts of most vitamins and minerals. . . . The manioc leaves, on the other hand, are rich in protein, calcium, and ascorbic acid, and contain significant amounts of iron and of the A and B vitamins. . . . But when the leaves as well as the roots are eaten, manioc comes close to justifying the name of 'the all sufficient' that was given it by natives in southwestern Congo because 'We get bread from the root and meat from the Leaves.'"[10] In many places where manioc is cultivated in Africa, the leaves are cooked and served as spinach or added to sauces as a condiment. Manioc is an example of African culinary innovation with an introduced plant. It is evidence of the ways that longstanding dietary preferences and foodways have guided the adoption of new crops.

The culinary emphasis on leafy vegetables is similarly evident in the cuisines of survival that Africans developed in New World plantation societies. Some greens apparently came to the Americas with enslaved Africans, notably mustard greens and collards. The African eggplant or guinea squash, like sesame and taro, were also cultivated for their edible leaves.[11] Greens were used both as food and as medicine. Sorrel or hibiscus leaves, for instance, are prized as a cooked green, and parts of its flower are made into a beverage; decoctions of the dried roots, leaves, and seeds are important curatives in West African and New World diasporic communities.[12]

Among the edible greens important in diaspora foodways are species that belong to plant genera distributed on both sides of the Atlantic. One is "bitter leaf" (*Vernonia* spp.), a plant widely used in West African cooking for the taste it imparts to broths and its reputed medicinal properties.[13] The *Amaranthus* genus also includes several species that are consumed as boiled spinaches by Africans and diasporic populations alike.[14] The significance of vegetable amaranths in the cooking traditions of Africans throughout the Atlantic world suggests that slaves in the Americas recognized similar members of the genus and substituted a New World species for a familiar African equivalent. The replacement of an African amaranth with a New World cousin is captured in language. Francis Moore, a slave trader who worked along the Gambia River in the early eighteenth century, wrote of one cultivated green, "an Herb call'd Colliloo, much like Spinage," which he claimed "eats almost as well."[15] Across the Atlantic in Jamaica, the African word *callalou* refers to spinach made from the leaves of New World amaranth species.[16]

In many plantation societies, the word *callalou* named not only the greens of the *Amaranthus* genus but also the soups and stews that featured them. Eighteenth-century missionary C. G. A. Oldendorp encountered these stews in the Danish Virgin Islands and borrowed the word *calelu* from the African slaves who prepared and ate them.[17] Over time, enslaved women and their descendants working in plantation kitchens transformed this stew into the more richly embellished *callalou* of our time. Today *callalou* is recognized as one of the signature dishes of the African diaspora. These flavorful pepper pots of the circum-Caribbean bring together foods from different cultural traditions— African, Amerindian, and European—into a single dish, still known as *callaloo* (or *callalou; caruru* in Brazil).[18] An important variant of this dish is the *gumbo,* in which okra substitutes for greens as the principal vegetable. Both *callalou* and *gumbo* are African words for African ingredients that lend the dishes their culinary definition. Each expresses the African preference for greens and the continuity of African cooking practices in the Americas.[19]

FOUNDATIONS OF DIASPORA FOODWAYS

Callalou and *gumbo* are today typically served over rice. But in the plantation era, rice cultivation was limited to a few New World regions and so the cereal was not common fare for most slaves. Where rice could not serve as the starchy base of a meal, diaspora dishes more closely resembled those of African food traditions that are instead based on foodways that use rain-fed cereals or tubers. Willem Bosman, the Dutch factor at the slave fort of Elmina from 1688 to 1702, provided an early description of the two basic meals Africans consumed: "Their common Food is a Pot full of ground Millet or Corn boiled to the consistence of Bread or instead of that Jambs [yams] and Potatoes; over which they pour a little Palm-Oyl, with boiled Vegetables and a little piece of stinking Fish."[20] Bosman's commentary is one of the earliest European comparisons of two fundamental foodways of the Guinea coast, in which different dietary starches underlie each dish. One builds on a cereal (millet, or "corn"), the other on a tuber. The distinction derives from staple preferences that predominate in different African environments: cereals typically form the basis of meals in savanna areas, tubers in humid tropical regions receiving abundant rainfall. Each typifies longstanding culinary traditions of western Africa.

In more equatorial climates, tubers are the predominant staple and thus often replace cereals in local cooking traditions. The meal Bosman described being made with yams or potatoes is the celebrated *fufu* of Ghana, Nigeria, and Cameroon. *Fufu* is made by boiling starchy tubers (plantain and taro are also used) in water until they are soft and then pounding them into a pulp. Continuous stirring in a large mortar causes the glutinous mass to reach a sticky consistency with an appearance similar to, but much thicker than, mashed potatoes. It is then typically spooned into individual servings that are garnished with seasonings, vegetables, and other ingredients. Another way to serve *fufu* is to mix the toppings directly into the mash.

Fufu became important in New World plantation societies where the main carbohydrate derived from tubers.[21] The dish was known in Brazil as *angú*. The early nineteenth-century artist Jean Baptiste Debret depicted a group of female vendors stirring kettles of manioc-based *angú*, which he wrote was served with okra, greens, and other side dishes (figure 10.1).[22] In many parts of the Caribbean, *fufu* is made with plantains, just as it is in western Africa. The same dish is known in the Dominican Republic as *mangú*, in Puerto Rico as *mofongo*, and in Cuba as *fufu de plátanos*. In each country the emphasis on greens served with a starchy base and prepared in the African way persists.

Bosman also noted the use of cereals as a foundation for an African meal. These "porridges" were traditionally made from sorghum, millet, or fonio but later included maize after its introduction. In the cooking traditions of other cultures, the equivalent starchy formulations would be known as polenta, dumplings, or cornbread. African porridge is prepared by placing unhusked grain in an upright wooden mortar, where it is pounded by hand with a pestle into flour. The flour is gradually added to boiling water until the mixture reaches a thick consistency. The spongy "bread" is then served in bowls to which beans, okra, leafy vegetables, and other ingredients are added. In an alternate version, the porridge is formed into dumplings and dipped into side dishes. A more

FIGURE 10.1. *Negresses marchandes d'angou* (Black women sellers of *angú*), Brazil, by Jean Baptiste Debret, ca. 1821–25.

SOURCE: Debret, *Viagem pitoresca e histórica ao Brasil,* vol. 1, pl. 35, following p. 212.

elaborate preparation, known as *kenkey,* is made by allowing partially cooked porridge to ferment. The sour dough is then divided into dumplings, which are wrapped with plantain leaves or corn husks and steamed. An illustration accompanying Pieter de Marees' description of the Gold Coast at the end of the sixteenth century depicts African women selling *kenkey (kanquies)* in a market frequented by the Dutch (see figure 3.1).[23]

As chattel in the slave ports of Guinea and as prisoners on slave ships, Africans received degraded versions of these basic meals, which had been reduced by slavers to an insubstantial gruel. This gruel became the porridge, mush, or pap of slaveholder accounts. It was left to African women in the New World to restore these diluted staples to familiar African formulations. Several ingredients and dishes of the diaspora bear African names. Some cereal-based porridges of Africa, for example, were known in the Americas as *fundi, funchi,* or *funji.* The words likely derive from *nfundi,* the name reported for the basic porridge of seventeenth-century Kongo, which was served with sauces, greens, and side dishes.[24] References to *fundi* also appear in eighteenth-century records from the West Indies. Dutch documents mention that the "staple food served to slaves at Curaçao was small maize [sorghum]. . . . Small maize was ground into flour and then boiled into porridge (called *funchi*) or baked into cakes."[25] In the Danish Virgin Islands, Oldendorp described "the everyday food of the Negroes" as maize, cooked into mush or "baked into small cakes or prepared as funji, a kind of large dumpling which is eaten with calelu."[26]

The other foundation of diaspora cooking was rice. In rice-growing regions such as South Carolina, Louisiana, Maranhão (Brazil), and the Guianas, maroon and enslaved females innovated with a number of rice-based dishes, some prepared with greens, others with beans of African origin.[27] A regional favorite of Maranhão is *arroz de cuxá*, or rice cooked with sorrel leaves (*Hibiscus sabdariffa*, called *vinagreira* in Portuguese). *Cuxá* is undoubtedly a loan word from *kucha*, the Mandinka name for African sorrel in the rice-growing region of Senegambia.[28]

Hoppin' John, made of rice and black-eyed peas, is a quintessential dish of the southern United States—in particular, South Carolina. It was already a long-established lowcountry favorite when Sarah Rutledge, the daughter of a prominent Charleston slaveholding family, published the recipe in her 1847 cookbook, *The Carolina Housewife*.[29] Although Hoppin' John is a Southern dish, its contours are distinctly African, with two main African ingredients and origins linked to the slave dwellings and plantation kitchens of the South. Hoppin' John is traditionally prepared alongside plates of collard greens for New Year's Day and is reputed to bring good luck to all who have it as their first meal of the year.

MARKET WOMEN AND THE
DIFFUSION OF AFRICAN FOODS

As vendors of prepared food, "market women" also promoted a wider acceptance of diaspora cuisines among New World populations (figure 10.2). Cooking and selling food were common occupations of enslaved and free females, much as it was for African women in Guinea's traditional markets. Female vendors were especially active in the sale of fresh vegetables and fruits. The banana leaf conspicuously draped across the woman's fruit tray in plate 8 connotes the leaf's importance in eighteenth-century cooking practices of Brazil. The banana or plantain leaf provided a wrapper for steamed food in much the same way that the corn husk encloses the tamale in indigenous Meso-American foodways. However, in this image the culinary inflection is clearly African rather than Amerindian. A well-known exemplar of this cooking tradition is *abará*, a Bahian dish of steamed black-eyed pea meal wrapped in banana leaf.

FIGURE 10.2. *A Free Negress and Other Market-Women,* Rio de Janeiro, by James Henderson, 1821.

SOURCE: Henderson, *History of the Brazil,* opposite p. 71.

The market women of plantation societies were variously known as "higglers" and "hucksters" in the British Caribbean and as *quitandeiras* in Brazil.[30] These female vendors also specialized in selling prepared beverages and cooked food (figure 10.3). Through these activities, market women perpetuated African and diaspora convenience foods and drinks. In tropical America, two African beverages are especially popular for their refreshing tastes: *tamarindo,* made from the pulp of the tamarind pod; and a tart, cranberry-like drink, made from the sepals of the hibiscus flower *(Hibiscus sabdariffa),* known as *flor de Jamaica* in Spanish-speaking America.

Notable among the common convenience foods made by enslaved and free women were fritters. A fritter could consist of fish, vegetables, fruits, rice, or cornmeal deep fried in vegetable oil. One variation is the hushpuppy, a Southern favorite made from cornmeal; another is bean cake prepared from black-eyed peas. The latter was sold by market women in South Carolina well into the twentieth century and remains popular to this day in Brazil as *acarajé.* Both kinds of fritter are made by pounding the main ingredient with a pestle, forming a patty or a ball with the hands, and then deep-frying it in vegetable oil. This West African cooking practice was observed in the fourteenth century. On his journey to Mali in 1352, Muslim traveler Ibn Battuta mentioned fritters

FIGURE 10.3. *The Quitandeira* (The market woman), 1857.

SOURCE: Daniel P. Kidder, *Brazil and the Brazilians, Portrayed in Historical and Descriptive sketches* (Philadelphia, 1857), 167, in Handler and Tuite, "Atlantic Slave Trade and Slave Life in the Americas: A Visual Record," database at University of Virginia Library, image ref.: Kidder 7.

made from indigenous Bambara groundnuts that were fried in *karité* (the French word for the vegetable butter of the Sahelian shea nut tree).[31] The vegetable oil of choice for fritters in Africa's humid tropics comes from the oil palm. It remains a favorite cooking oil of Afro-Bahian cooking, where it is known as *dendê* (from the Kimbundu word *ndende*).[32] Deep-frying with vegetable oil is an ancient cooking tradition in West Africa that enslaved African women likely introduced to plantation societies.[33]

FORCED MIGRATION AND MEMORY

The journey of humble foods such as the fritter from West Africa to New World plantation societies to our own place and time speaks to food's importance as a touchstone of the experience of human migration. Migrants the world over bring their dietary preferences and cooking practices with them. These traditions are rarely forsaken, even when food preferences cannot be reconstituted in full. Food gives material expression to the ways exiles commemorate the past and shape new identities amid alien cultures, diets, and languages.[34] Food is vested with symbolic ties to homelands left or lost. The emphasis on meaningful foods and familiar forms of preparation enriches the memory dishes with which migrants connect past and present.

No less than other immigrant groups who came to the New World, enslaved Africans also arrived with specific food preferences and cooking traditions. Few slaves were able to complete the Middle Passage with anything more tangible than their memories. Yet the victualing demands of the Atlantic crossing ensured that Africa's principal dietary staples were also frequent travelers on slave ships. In the early colonial period, these staples found a New World footing in the spaces slaves cultivated around their dwellings. The violent dislocations of New World bondage could not eradicate the memories of African foodways, and indeed invigorated and perpetuated them because of their importance as means of subsistence and survival.

Food was central to the experience of having been made a chattel slave. For trade slaves bound for the Americas, bondage severed not least the right to partake in the customary foodways that affirmed membership in an African culture. The rupture was reinforced by the degradations of the Middle Passage, and it persisted in the subsistence regimes of New World slave societies, where the enslaved were fed meager rations or left to fend for themselves. Survival depended critically on the extra exertions slaves made to diversify and augment basic needs. But just as importantly, these extra exertions reconstituted and renewed some customary African foodways. Perhaps in no small part it is the cultural memory of slavery and hunger that sometimes makes food, especially food of African origin, a metaphor for migration and loss among diasporic cultures.[35]

Out of the exigencies of food, diasporic peoples vested many African staples with important symbolic meanings. The annual *seú* festival of Curaçao, for instance, began as a celebration of the sorghum harvest; in Suriname, yams are prepared for the ancestral offerings made by slave-descended populations; and the New Year's Day dish of rice and black-eyed peas served in the South since slavery times betokens good fortune. Rice, in particular, is foundational to the commemorative dishes of many Maroon societies. Other African foods are prominently featured in the liturgical offerings of Afrosyncretic religious practitioners throughout the Americas. The black-eyed pea dishes *abará* and *acarajé,* for example, are both among the consecrated specialties of candomblé cooking.[36]

These are all celebrations of food, and it is through the lens of mere food— humble foods—rather than plantation commodities, that we begin to understand the role of Africans in shaping New World farming systems, animal-husbandry practices, and regional cuisines. The marginal spaces of slave food plots offer more insight into this neglected history than the estate fields where they toiled.

African contributions to the global table began with the journeys of several crops across time and space. Nowhere was African agency more transformative than in the oppressive landscapes of New World slavery. The complex relationships of plants and people that were sundered by the Middle Passage were quietly recast by enslaved Africans and their descendants in the food fields and kitchens of plantation societies. What distinguishes the foodways of the African diaspora are their humble beginnings and the discrete ways they infiltrated the plantation palate. What makes them remarkable is the story they tell of exile, survival, endurance, and memory.

NOTES

INTRODUCTION

Epigraph: Sidney W. Mintz, review of *Black Rice,* by Judith A. Carney, *Annals of the Association of American Geographers* 92, no. 4 (2002): 799.

1. Alfred W. Crosby, *The Columbian Exchange: Biological and Cultural Consequences of 1492* (Westport, CT: Greenwood Press, 1972).

2. Alfred W. Crosby, *Ecological Imperialism: The Biological Expansion of Europe, 900–1900* (New York: Cambridge University Press, 1986).

1. FOOD AND THE AFRICAN PAST

Epigraphs: The Akan proverb is one heard many times by the authors; the Bassari legend is quoted in Joseph Campbell, *Historical Atlas of World Mythology,* 3 vols. (New York: Harper & Row, 1988), vol. 1, part 1, p. 14. The Akan people are an important ethnic group in Ghana. They have been associated with gold mining and the long-distance trade in gold since the Middle Ages. During the sixteenth century, they formed the powerful, centralized Ashanti kingdom, which strongly resisted political control until the twentieth century. The Bassari are a people who live in the Futa Jallon plateau of Guinea. They are also found in the eastern parts of Senegambia and Guinea-Bissau.

1. B.C.E. ("before the common era," which is year 1 of the Gregorian calendar) replaces B.C. in this discussion. B.P. ("before the present") is used for long-term climatic and geological processes, but interchangeably means "years ago."

2. National Research Council (NRC), *Lost Crops of Africa,* vol. 1, *Grains* (Wash-

ington, D.C.: National Academy Press, 1996); NRC, *Lost Crops of Africa,* vol. 2, *Vegetables* (Washington, D.C.: National Academy Press, 2006).

3. These are, respectively, the kola nut, oil palm, tamarind, hibiscus, and gum arabic.

4. Francisco de Lemos Coelho traded along the Gambia River in the late 1650s and wrote an account that was completed in 1669. David P. Gamble and P. E. H. Hair, eds., "Francisco De Lemos Coelho," in *The Discovery of River Gambra (1623) by Richard Jobson* (London: Hakluyt Society, 1999), 299–300.

5. Scholars continue to debate the precise timing of the development of full language, but an important recent study places the emergence of a human vocal tract capable of articulating human speech after 100,000 B.P. and before 50,000 B.P. See Philip Lieberman and Robert McCarthy, "Tracking the Evolution of Language and Speech: Comparing Vocal Tracts to Identify Speech Capabilities," *Expedition* 49, no. 2 (Summer 2007): 15–20.

6. Anatomically modern humans penetrated the Levant as early as 120,000–100,000 years ago, but as Lieberman and McCarthy show, these anatomically modern humans did not yet have the modern vocal tract; in other words, they had not yet fully evolved into *Homo sapiens sapiens* like ourselves. These early anatomically modern humans did not persist in the Levant but apparently died out after a new advance of *Homo neanderthalensis* populations into the region. The Neanderthal immigrants were better adapted to the colder conditions that began some 70,000 years ago. Lieberman and McCarthy, "Tracking the Evolution of Language and Speech."

7. R. Marshall and E. Hildebrand, "Cattle before Crops: The Beginnings of Food Production in Africa," *Journal of World Prehistory* 16, no. 2 (2002): 100.

8. Rudolf Kuper and Stefan Kroepelin, "Climate-Controlled Holocene Occupation in the Sahara: Motor of Africa's Evolution," *Science* 313, no. 5788 (2006): 803–7.

9. Fred Wendorf and Romuald Schild, "Nabta Playa and Its Role in Northeastern African Prehistory," *Journal of Anthropological Archaeology* 17 (1998): 97–123; Kuper and Kroepelin, "Climate-Controlled Holocene," 804.

10. The period following the last Pleistocene ice age (from 11,500 B.P.) is known as the Holocene. Human civilizations across the world made considerable adaptations to global climatic shifts, including plant and animal domestication.

11. Humpless taurine cattle were formerly thought to have been domesticated in the Fertile Crescent, but DNA analyses of African breeds now suggest an independent, not derivative, population in the region from southern Libya to southern Egypt. D. G. Bradley, D. E. MacHugh, P. Cunningham, and R. T. Loftus, "Mitochondrial Diversity and the Origins of African and European Cattle," *Proceedings of the National Academy of Sciences (PNAS)* 93, no. 10 (1996): 5131–35; Olivier Hanotte, D. G. Bradley, J. W. Ochieng, Y. Verjee, E. W. Hill, and J. E. Rege, "African Pastoralism: Genetic Imprints of Origins and Migrations," *Science* 296, no. 5566 (2002): 336–39; Ciaran Meghen, David MacHugh, B. Sauveroche, G. Kana, and Dan Bradley, "Characterization of the Kuri Cattle of Lake Chad Using Molecular Genetics Techniques," in Roger M. Blench and Kevin C. MacDonald, eds., *The Origins and Development of African Livestock:*

Archaeology, Genetics, Linguistics and Ethnography (London: University College London, 2000), 259–68; Wendorf and Schild, "Nabta Playa"; Marshall and Hildebrand, "Cattle before Crops," 113.

12. The Red Sea Hills region extends from the northern edge of the Ethiopian Highlands through Sudan to the eastern part of Egypt.

13. Alfred Muzzolini, *L'Art rupestre préhistorique des massifs centraux sahariens* (Oxford: British Archaeological Reports, 1986); Augustin F. C. Holl, *Saharan Rock Art: Archaeology of Tassilian Pastoralist Iconography* (New York: Altamira Press, 2004), 7, 11.

14. Kuper and Kroepelin, "Climate-Controlled Holocene."

15. Marshall and Hildebrand, "Cattle before Crops."

16. The *n'dama* breed is also known as Mandingo cattle because of its association with the language group in the modern country known as Guinea. The small breed weighs less than other cattle, typically no more than 250–350 kilograms. It is also markedly resistant to tick-borne infections. G. Williamson and W. J. A. Payne, *An Introduction to Animal Husbandry in the Tropics* (London: Longmans, 1960), 138–44, 155.

17. H. Epstein, *The Origin of the Domestic Animals of Africa*, 2 vols. (New York: Africana Publishing, 1971), 1:201–326, esp. 202–3.

18. Juliet Clutton-Brock, *A Natural History of Domesticated Mammals* (Cambridge: Cambridge University Press, 1999), 114–26; Stine Rossel, Fiona Marshall, Joris Peters, Tom Pilgram, Matthew D. Adams, and David O'Connor, "Domestication of the Donkey: Timing, Processes, and Indications, *PNAS* 105, no. 10 (2008): 3715–20; Kevin C. MacDonald, "The Origins of African Livestock: Indigenous or Imported?" in Blench and MacDonald, *Origins and Development of African Livestock*, 2–17, esp. 10–11. Linguistic studies suggest that donkeys were domesticated by Afro-Asiatic language speakers. Christopher Ehret, "Historical/Linguistic Evidence for Early African Food Production," in J. Desmond Clark and Steven A. Brandt, eds., *From Hunters to Farmer: The Causes and Consequences of Food Production in Africa* (Berkeley: University of California Press, 1984), 27.

19. There are four genera of guinea fowl in Africa, but for domestication *Numida meleagris* is the most significant. The West African subspecies *N. Meleagris galeata* was introduced to New World plantation societies at an early date. R. A. Donkin, *Meleagrides: An Historical and Ethnogeographical Study of The Guinea Fowl* (London: Ethnographica, 1991).

20. Epstein, *Origin of the Domestic Animals of Africa*, 2:21–56.

21. Scientists call this fundamental selection process the "domestication syndrome." Melinda A. Zeder, Daniel G. Bradley, Eve Emshwiller, Bruce D. Smith, "Documenting Domestication: Bringing Together Plants, Animals, Archaeology, and Genetics," in M. Zeder, D. Bradley, E. Emshwiller, B. Smith, eds., *Documenting Domestication: New Genetic and Archaeological Paradigms* (Berkeley: University of California Press, 2006), 1–12; P. Gepts, "A Comparison between Crop Domestication, Classical Plant Breeding, and Genetic Engineering," *Crop Science* 42 (2002): 1781.

22. Sorghum that was morphologically wild was harvested in the Sahara before 8,000 years ago. Fred Wendorf, Angela Close, Romuald Schild, Krystyna Wasylikowa, Rupert Housley, Jack R. Harlan, and Halina Królik, "Saharan Exploitation of Plants 8,000 years BP," *Nature* 359, no. 6397 (1992): 721–24. But the presence of bell-shaped pit granaries at archaeological sites suggests that this was cultivated sorghum, although ongoing interbreeding with wild sorghum had not yet made it morphologically distinct. Christopher Ehret, Department of History, University of California, Los Angeles, pers. comm. December 24, 2008. The older date for domesticated sorghum derives from a site excavated in the 1970s. Anne Brower Stahl, "Early Food Production in West Africa: Rethinking the Role of the Kintampo Culture," *Current Anthropology* 27, no. 5 (1986): 532–36; Jack R. Harlan, "The Tropical African Cereals," in David R. Harris and Gordon C. Hillman, eds., *Foraging and Farming: The Evolution of Plant Exploitation* (London: Unwin Hyman, 1989), 335–43, esp. 337.

23. Lech Krzyzaniak, "New Light on Early Food-Production in the Central Sudan," *Journal of African History* 19, no. 2 (1978): 159–72; Christopher Ehret, "Sudanic Civilization," in Michael Adas, ed., *Agricultural and Pastoral Societies in Ancient and Classical History* (Philadelphia: Temple University Press, 2001), 224–74.

24. Christopher Ehret, *An African Classical Age: Eastern and Southern Africa in World History, 1000 B.C. to A.D. 400* (Charlottesville: University Press of Virginia, 1998), 14.

25. Linguistics research supports the potential significance of West Africa for early plant domestication on the continent. Christopher Ehret, "Linguistic Stratigraphies and Holocene History in Northeastern Africa," in Marek Chlodnicki and Karla Kroeper, eds., *Archaeology of Early Northeastern Africa,* Studies in African Archaeology 9 (Posnán, Poland: Posnán Archaeological Museum, 2006), 1019–55.

26. Jack R. Harlan, Jan M. J. de Wet, and Ann B. L. Stemler, *Origins of African Plant Domestication* (The Hague: Mouton, 1976); F. R. Irvine, "Supplementary and Emergency Food Plants of West Africa," *Economic Botany* 6 (1952): 23–40.

27. Roger M. Blench, *Archaeology, Language, and the African Past* (Lanham, MD: Altamira Press, 2006), 211–12, 257–58; Katharina Neumann, "Early Plant Food Production in the West African Sahel: New Evidence," in Marijke Van der Veen, ed., *The Exploitation of Plant Resources in Ancient Africa* (New York: Kluwer Academic, 1999), 73–80, esp. 85.

28. Jack R. Harlan, "Agricultural Origins: Centers and Noncenters," *Science*, New Series 174, no. 4008 (1971): 468–74.

29. NRC, *Lost Crops*, 1:128; Harlan, "Tropical African Cereals"; Jan M. J. de Wet, "Sorghum," in Kiple and Ornelas, *Cambridge World History of Food*, 1:152–58.

30. Marshall and Hildebrand, "Cattle before Crops," 130–31; Reader, *Africa: A Biography*, 251.

31. William S. Pollitzer, *The Gullah People and Their African Heritage* (Athens: University of Georgia Press, 1999), 95–96.

32. Enset is cultivated for the lower trunk's starchy pith and the underground edi-

ble corm, the part that extends up to its lower stem. Lejju et al., "Africa's Earliest Bananas?"; NRC, *Lost Crops*, 2:173–89. Historical linguistics of the Nilo-Saharan language family support the antiquity of many of these plants in the southern eastern Sahara between 8500 and the sixth millennium B.C.E. Ehret, "Linguistic Stratigraphies."

33. Finger millet is one of the world's most nutritious cereals. Some varieties have high levels of methionine, an amino acid missing from starchy staples such as cassava and plantain that dominate the diets of millions of the world's poor. NRC, *Lost Crops*, 1:39.

34. Pearl millet and guinea millet *(Brachiaria deflexa)* have been found in an ancient lake bed in Mali, in an archaeological site that dates to the middle of the second millennium B.C.E. Stahl, "Early Food Production"; A. C. D'Andrea and J. Casey, "Pearl Millet and Kintampo Subsistence," *African Archaeobotanical Review* 19, no. 3 (2002): 147–73.

35. NRC, *Lost Crops*, 1:79–81.

36. These are white- and black-grained, respectively.

37. Tadeusz Lewicki, *West African Food in the Middle Ages* (Cambridge: Cambridge University Press, 1974), 35; NRC, *Lost Crops*, 1:59–75.

38. Susan K. McIntosh, "Paleobotanical and Human Osteological Remains," in S. K. McIntosh, ed., *Excavations at Jenné Jeno, Hambarketolo, and Kaniana (Inland Niger Delta, Mali, the 1981 Season)* (Berkeley: University of California Press, 1995), 348–53.

39. Duncan A. Vaughan, Bao-Rong Lu, and Norihiko Tomooka, "The Evolving Story of Rice Evolution," *Plant Science* 174 (2008): 394–408; Roland Portères, "African Cereals: Eleusine, Fonio, Black Fonio, Teff, Brachiaria, Paspalum, Pennisetum, and African Rice," in Jack R. Harlan, Jan M.J. de Wet, Ann B.L. Stemler, eds., *Origins of African Plant Domestication* (The Hague: Mouton, 1976), 409–452.

40. Christopher Ehret, Department of History, University of California, Los Angeles, pers. comm., December 24, 2008.

41. Judith A. Carney, *Black Rice: The African Origins of Rice Cultivation in the Americas* (Cambridge, MA: Harvard University Press, 2001), 38–39.

42. The African eggplant was the first of two related species that Europeans came across. It was named because the size, shape, and color of the vegetable resembled a hen's egg. The purple Asian species retained the original, but inappropriate, African name when it became more popular in global foodways. The African eggplant is the guinea squash reported in plantation societies. It is esteemed for its bitter taste and edible green leaves. NRC, *Lost Crops*, 2:137–53.

43. *Egusi* often refers to the seeds of watermelon or another bitter melon, also called *egusi (Cucumeropsis edulis).*

44. The fluted pumpkin *(Telfairia occidentalis)* has also long been cultivated for its protein-rich seeds. Daniel Zohary and Maria Hopf, *Domestication of Plants in the Old World* (Oxford: Oxford University Press, 2000), 193–94; NRC, *Lost Crops*, 11:158; Irvine, "Supplementary and Emergency Food Plants."

45. But on the plant's potential Asian origin, see D. L. Erickson, Bruce D. Smith, A. C. Clarke, D. H. Sandweiss, and N. Tuross, "An Asian Origin for a 10,000-Year-Old Domesticated Plant in the Americas," *PNAS* 102, no. 51 (2005): 18315–20. The bot-

tleneck gourd may have reached the Americas via Atlantic drift as early as 9000 B.C.E., making its domestication date even earlier than other African crops. Thomas J. Riley, Richard Edging, and Jack Rosen, "Cultigens in Prehistoric Eastern North America: Changing Paradigms," *Current Anthropology* 31, no. 5 (1990): 525–41; Deena S. Decker-Walters, Mary Wilkins-Ellert, Sang-Min Chung, and Jack E. Staub, "Discovery and Genetic Assessment of Wild Bottle Gourd [*Lagenaria Siceraria* (Mol.) Standley; Cucurbitaceae] from Zimbabwe," *Economic Botany* 58, no. 4 (2004): 501–8; Peter H. Wood, "The Calabash Estate: Gourds in African American Life and Thought," in *African Impact on the Material Culture of the Americas* (Winston-Salem, NC: Museum of Early Southern Decorative Arts, 1998).

46. Dorothea Bedigian, "Sesame in Africa: Origin and Dispersals," in Neumann et al., *Food, Fuel and Fields*, 17–36; Dorothea Bedigian, pers. comm., August 23, 2009; Zohary and Hopf, *Domestication of Plants*.

47. *Momordica* is also known in popular parlance as balsam apple. Hibiscus gives Red Zinger tea its color.

48. A. M. G. Rutten, *Dutch Transatlantic Medicine Trade in the Eighteenth Century under the Cover of the West India Company* (Rotterdam: Erasmus, 2000), 103–4.

49. H. M. Burkill, *The Useful Plants of West Tropical Africa*, 6 vols. (Kew, England: Royal Botanic Gardens, 1985–2004), 2:216–302.

50. However, the pigeon pea may have reached Africa in ancient times. If it did, then the West African humid tropics were an important secondary center of diversification.

51. The *Vigna* genus originated in tropical Africa, *Phaseolus* in Central and South America. Only in the Americas do the words "bean" and "pea" differentiate the two types. This is likely because the small size of the former made them seem like the green peas that Europeans cultivated. NRC, *Lost Crops*, 1: 105–6, 223–24. Legumes the world over are esteemed for improving soil fertility because they trap atmospheric nitrogen in nodules on their roots, making it available to the soil.

52. A pantropical crop, there are also New World *(Dioscorea trifida)* and Asian yam species *(D. alata)*, of which the Asian is also an important tropical foodstaple. The yam is recognized as an extremely effective plant substitute for estrogen during female menopause. D. G. Coursey and C. K. Coursey, "The New Yams Festivals of West Africa," *Anthropos* 66, nos. 3–4 (1971): 444–84; J. Pérez, D. Albert, S. Rosete, L. Sotolongo, M. Fernandez, P. Delprete, and L. Raz, "Consideraciones etnobotánicas sobre el género Dioscorea *(Dioscoreaceae)* en Cuba," *Ecosistemas*, no. 2 (2005): 1–8; L. Brydon, "Rice, Yams and Chiefs in Avatime: Speculations on the Development of a Social Order," *Africa* 51, no. 2 (1981): 659–77.

53. Reader, *Africa: A Biography*, 251.

2. AFRICAN PLANTS ON THE MOVE

Epigraphs: The Akan proverb is one heard many times by the authors; the Bassari legend is quoted in Joseph Campbell, *Historical Atlas of World Mythology*, 3 vols. (New York: Harper & Row, 1988), vol. 1, part 1, p. 14.

1. Hannibal used the African forest elephant *(Loxodonta cyclotis)*, a smaller and separate species (as DNA testing now reveals), between 7 and 8 feet tall, not the great bush elephant *(L. africana)* of the African savannas, standing 11–12 feet at the shoulder. The forest elephant, about the size of a horse, was not yet extinct in North Africa in the third century B.C.E., when Hannibal initiated his expedition in Spain. Gavin de Beer, *Hannibal: Challenging Rome's Supremacy* (New York: Viking, 1969), 100–107.

2. Principal crops that likely were exchanged in ancient times between Asia and the New World include the sweet potato, banana, coconut, ginger, and possibly amaranth and cotton.

3. In *The Histories* (written between 431 and 425 B.C.E.), Herodotus also mentioned claims that Phoenician sailors "saw the sun on the right side" on their return journey from Africa. This mention suggests that ancient seafarers familiar with the eastern African coast may indeed have rounded the Cape of Good Hope and returned northward to the Mediterranean via West Africa. If they had circumnavigated the continent, the morning sun would have been on their right for the return journey northward.

4. This overview is based on the archaeological research team led by Professors David Mattingly and Andrew Wilson. The research findings are summarized in David Keys, "Kingdom of the Sands," *Archaeology* (March/April 2005): 26–29; and Ruth Pelling, "Garamantian Agriculture and Its Significance in a Wider North African Context: The Evidence of the Plant Remains from the Fazzan Project," *Journal of North African Studies* 10, no. 3 (2005): 397–411, esp. 398. See also E.W. Bovill, *The Golden Trade of the Moors* (Oxford: Oxford University Press, 1958), 21; and Henri Lhote, *The Search for the Tassili Frescoes* (New York: E.P. Dutton, 1959), 129–30.

5. A.I. Wilson and D.J. Mattingly, "Irrigation Technologies: Foggaras, Wells and Field Systems," in D.J. Mattingly, C.M. Daniels, J.N. Dore, D. Edwards, and J. Hawthorn, eds., *The Archaeology of Fazzan*, vol. 1, *Synthesis* (London: Society for Libyan Studies, 2003), 238–41.

6. Keys, "Kingdom," 25.

7. On the timeline for food crop introductions in the archaeological record, see Pelling, "Garamantian Agriculture," 401.

8. Keys, "Kingdom," 27. In 2007 the BBC reported the "discovery" in the region of an ancient underwater lake the size of Lake Erie. Now completely vanished, it was located underneath Libya, Egypt, and Sudan.

9. Alexander Popovic, *The Revolt of African Slaves in Iraq in the 3rd/9th Century* (Princeton, NJ: Markus Wiener, 1999), 11.

10. St. Clair Drake, *Black Folk Here and There*, 2 vols. (Los Angeles: UCLA/CAAS, 1990), 2:117–21; Bernard Lewis, *Race and Color in Islam* (New York: Harper and Row, 1971), 66.

11. Gregory Dicum and Nina Luttinger, *The Coffee Book: Anatomy of an Industry from Crop to the Last Drop* (New York: The New Press, 1999); Maguelonne Toussaint-Samat, *History of Food* (Malden, MA: Blackwell, 1999), 595; Paul E. Lovejoy, *Caravans of Kola: The Hausa Kola Trade, 1700–1900* (Zaria, Nigeria: Ahmadu Bello University

Press, 1980); A.M.G. Rutten, *Dutch Transatlantic Medicine Trade in the Eighteenth Century under the Cover of the West India Company* (Rotterdam: Erasmus, 2000), 90.

12. Sorghum became a dietary staple of the Muslim commercial network, which integrated sub-Saharan African sources of gold with the Moorish empire, thereby setting up later Catholic Iberian quests for African gold. Sorghum could also substitute for millet in the preparation of a dietary staple, couscous. Ray A. Kea, "Expansions and Contractions: World-Historical Change and the Western Sudan World-System (1200/1000 B.C.–1200/1250 A.D.)," *Journal of World-Systems Research* 10, no. 3 (2004): 723–816; Yves Lacoste, *Ibn Khaldun: The Birth of History and the Past of the Third World* (London: Verso, 1984), 16–18, 30.

13. It is possible that sorghum made its way across trans-Saharan caravan routes to the Roman Empire even earlier via the Garamantes, who cultivated it along with African millet. The Roman historian Pliny described the introduction in his time (ca. 62–79 C.E.) of a black-grained millet to Italy, which Watson suggests may have been the African species. Alternatively, it might have been African *fonio,* a small-grained millet. Contact with Berbers certainly introduced ancient Rome to new cooking traditions and ingredients. Andrew M. Watson, *Agricultural Innovation in the Early Islamic World* (Cambridge: Cambridge University Press, 1983), 12. See also Lilia Zaouali, *Medieval Cuisine of the Islamic World* (Berkeley: University of California Press, 2007), 121.

14. Thomas F. Glick, *Islamic and Christian Spain in the Early Middle Ages* (Princeton: Princeton University Press, 1979), 81–82. In fact, polenta was frequently prepared from sorghum in northern Italy prior to the introduction of maize. Giovanni Rebora and Albert Sonnenfeld, *The Culture of the Fork* (New York: Columbia University Press, 2001), 125; Sophie D. Coe, *America's First Cuisines* (Austin: University of Texas Press, 1994), 15–16.

15. Dorian Q. Fuller, "African Crops in Prehistoric South Asia: A Critical Review," in Neumann et al., *Food, Fuel and Fields,* 239–71; Steven A. Weber, "Out of Africa: The Initial Impact of Millets in South Asia," *Current Anthropology* 39, no. 2 (1998): 267–74; Roger M. Blench, "The Movement of Cultivated Plants between Africa and India in Prehistory," in Neumann et al., *Food, Fuel and Fields,* 273–92; National Research Council (NRC), *Lost Crops of Africa,* vol. 1, *Grains* (Washington, D.C.: National Academy Press, 1996) 42, 127–30, 148, 182–84; Daniel Zohary and Maria Hopf, *Domestication of Plants in the Old World* (Oxford: Oxford University Press, 2000), 89.

16. B. Julius Lejju, P. Robertshaw, and D. Taylor, "Africa's Earliest Bananas?" *Journal of Archaeological Science* 33 (2006): 102–13.

17. The term distinguishes an earlier period of intercontinental plant exchanges prior to the one mediated by Europeans known as the Columbian Exchange. J.R. McNeill, "Biological Exchange and Biological Invasion in World History," paper presented at the 19th International Congress of the Historical Sciences, Oslo, August 6–13, 2000; Alfred W. Crosby, *The Columbian Exchange: Biological and Cultural Consequences of 1492* (Westport, CT: Greenwood Press, 1972).

18. Edmond de Langhe, *Banana and Plantain: The Earliest Fruit Crops?* Annual

Report (Montpellier, France: INIBAP, 1995), 6–8; Lejju et al., "Africa's Earliest Bananas?" 108. Recent genetics evidence increasingly suggests New Guinea as a separate center of banana domestication. Christopher Ehret, Department of History, University of California, Los Angeles, pers. comm., July 17, 2008.

19. de Langhe, *Banana and Plantain*, 8.

20. Edible types of bananas are thought to derive from two wild species, *Musa balbisiana* (BB) and *M. acuminata* (AA). Domestication added hybrids of the two species as well as triploid chromosome sets. We can divide edible bananas into five principal genomic types: AA, AAA, AB, AAB, and ABB. Only the AAB and AAA species are found in Africa, with the AAB cultivars known as plantains, the AAA as bananas (even though many cultivars are cooked). Kairn Kleiman, *The Pygmies Were Our Compass: Bantu and Batwa in the History of West Central Africa, Early Times to c. 1900 C.E.* (Portsmouth, NH: Heinemann, 2003), 96–99.

21. de Langhe, *Banana and Plantain*, 8.

22. Over 60 percent of the world's plantains are still produced and consumed in West and Central Africa. Phillip Rowe and Franklin E. Rosales, "Bananas and Plantains," in Jules Janick and James N. Moore, eds., *Fruit Breeding*, vol. 1, *Tree and Tropical Fruits* (New York: John Wiley, 1996), 167–211, esp. 167.

23. This claim and the phytolith evidence from Uganda suggest *Musa* cultivation around 2000 B.C.E.; the claims from Cameroon and Uganda are controversial. See Lejju et al., "Africa's Earliest Bananas?"; and Jan Vansina, "Bananas in Cameroun c. 500 BCE? Not Proven," *Azania* 38 (2004): 174–76.

24. C. M. Mbida, W. Van Neer, H. Doutrelepont, and L. Vrydaghs, "Evidence for Banana Cultivation and Animal Husbandry During the First Millennium BC in the Forest of Southern Cameroon," *Journal of Archaeological Science* 27 (2000): 151–62. But see Vansina, "Bananas in Cameroun."

25. John H. Parry, "Plantation and Provision Ground," *Revista de historia de America* 39 (1955): 1–20; Reader, *Africa: A Biography*, 301.

26. Edmond de Langhe, R. Swennen, and D. Vuylsteke, "Plantain in the Early Bantu World," *Azania*, 29–30 (1994–95): 147–60, esp. 156; de Langhe, *Banana and Plantain*.

27. The adoption of the "banana" is associated with the emergence of several kingdoms and large states during the first millennium C.E. On the banana in the Great Lakes area of East Africa, see D. L. Schoenbrun, "Cattle Herds and Banana Gardens," *African Archaeological Review* 11 (1993): 39–72. On plantain history in the Congo Basin and equatorial forest regions, see Jan Vansina, *Paths in the Rainforest* (Madison: University of Wisconsin, 1990).

28. Felipe Fernández-Armesto, *Before Columbus: Exploration and Colonization from the Mediterranean to the Atlantic, 1229–1492* (Philadelphia: University of Pennsylvania Press, 1987), 190–92.

29. Only two of the Madeira Islands have been inhabited: Madeira and Porto Santo.

30. Alfred W. Crosby, *Ecological Imperialism: The Biological Expansion of Europe, 900–1900* (New York: Cambridge University Press, 1986), 77.

31. However, the development of sugarcane plantations in the New World caused the Atlantic islands to lose their primacy in the first quarter of the sixteenth century.

32. Crosby, *Ecological Imperialism*, 78.

33. Alberto Vieira, "Sugar Islands: The Sugar Economy of Madeira and the Canaries, 1450–1650," in Schwartz, *Tropical Babylons*, 42–84, 57–58.

34. Vieira, "Sugar Islands," 42.

35. The Guanches are thought to have been related to the Berbers of the African mainland. Crosby, *Economic Imperialism*, 80–81.

36. These were the trading posts south of Cape Bojador on the African mainland. Linda A. Newson and Susie Minchin, *From Capture to Sale: The Portuguese Slave Trade to Spanish South America in the Early Seventeenth Century* (Boston: Brill, 2007), 2; Fernández-Armesto, *Before Columbus*, 192.

37. Crosby, *Ecological Imperialism*, 83–84, 96.

38. Rowe and Rosales, "Bananas and Plantains," 169; Vieira, "Sugar Islands," 68, 74; Charlotte Porter, "Science at the Time of Columbus," in James R. McGovern, ed., *The World of Columbus* (Macon, GA: Mercer University Press, 1992), 59–77, esp. 60.

39. According to the official Spanish historian Oviedo and mentioned in Crosby, *Ecological Imperialism*, 97.

40. Antonio Rumeu de Armas, *España en el Africa Atlántica*, 2 vols. (Madrid: Instituto de Estudios Africanos, Consejo Superior de Investigaciónes Científicas, 1956), 1:50–52, 56, 58, 115, 150.

41. Emilia Viotti da Costa, "The Portuguese-African Slave Trade: A Lesson in Colonialism," *Latin American Perspectives* 12 (1985): 41–61, esp. 50; Gerald Roe Crone, *The Voyages of Cadamosto and Other Documents on Western Africa in the Second Half of the Fifteenth Century* (London: Hakluyt Society, 1937), 18.

42. In the early period of Iberian overseas expansion, the Azenegues inhabited the area of Mauritania north of the Senegal River valley. They tilled the soil, collected gum arabic, and tended livestock herds. They had converted to Islam by the time the Portuguese established the slave post at Arguim. These "Moors" included Berbers and black Africans, who spoke a language now extinct. Because they lived in tributary relationships to their overlords, they were vulnerable to enslavement and being sold to slave traders operating out of Arguim.

43. The annual volume of the Portuguese trade in African slaves exceeded two thousand in the last decades of the fifteenth century. The Arguim fort repeatedly experienced food shortfalls. Ivana Elbl, "'Slaves Are a Very Risky Business . . .': Supply and Demand in the Early Atlantic Slave Trade," in José C. Curto and Paul E. Lovejoy, eds., *Enslaving Connections: Changing Cultures of Africa and Brazil during the Era of Slavery* (Amherst, NY: Humanity Books, 2004), 29–55. See also R. W. Unger, "Portuguese Shipbuilding and the Early Voyages to the Guinea Coast," in Felipe Fernández-Armesto, ed., *The European Opportunity*, vol. 2, *An Expanding World: The European Impact on World History, 1450–1800* (Ashgate, England: Variorum, 1995), 45; Linda M. Heywood

and John K. Thornton, *Central Africans, Atlantic Creoles, and the Foundation of the Americas, 1585–1660* (New York: Cambridge University Press, 2007), 68.

44. P. K. Reynolds, "Earliest Evidence of Banana Culture," *Journal of the American Oriental Society Supplement*, no. 12 (1951): 1–28; D. H. Marin, T. B. Sutton, and K. R. Barker, "Dissemination of Bananas in Latin America and the Caribbean and Its Relationship to the Occurrence of *Radopholus similis*," *Plant Disease* 82, no. 9 (1998): 964–74; N. W. Simmonds, *Bananas* (London: Longmans, 1959), 313.

45. The Spanish historian Oviedo mentions the introduction to Santo Domingo in his sixteenth-century account. See Anon., [ed.], "Origin of the Banana," *Journal of Heredity* 5, no. 6 (1914): 278. The Canary Islands are today an important producer of fruit bananas, the Cavendish type, which was probably introduced from Asia in the nineteenth century. They are also cultivated on Madeira. R. H. Stover and N. W. Simmonds, *Bananas* (London: Longmans, 1987), 114–16.

46. In the fifteenth-century treaty, the Spanish only recognized Portuguese claims south of Cape Bojador. Spain maintained its coastal enclaves in the desolate arid territory now known as the Western Sahara. This region today includes the southernmost extension of Morocco and the territory north of Mauritania, which were of strategic importance to Spain for their close proximity to the Canary Islands. Spain formalized control of the Western Sahara by claiming it as a colony until 1975. Ceded to Morocco, the area is still disputed in territorial claims between Morocco and the indigenous Sahrawi, backed by Algeria. A cease-fire in 1991 leaves the Western Sahara's independence unresolved.

47. Raleigh Ashlin Skelton, *Magellan's Voyage: A Narrative Account of the First Navigation by Antonio Pigafetta* (New York: Dover, 1994), 63.

48. Jean Barbot, "A Description of the Coasts of North and South Guinea; and of Ethiopia Inferior, Vulgarly Angola; Being a New and Accurate Account of the Western Maritime Countries of Africa," in Churchill, *Collection of Voyages and Travels*, vol. 5, pl. F, p. 128.

49. Reynolds, "Earliest Evidence," 26.

50. Lopez's original description of western Africa was mentioned by Felipe Pigafetta in 1591. P. K. Reynolds, *The Banana: Its History, Cultivation and Place among Staple Foods* (New York: Houghton Mifflin, 1927), 27.

51. Fernandes, quoted in Stanley B. Alpern, "The European Introduction of Crops into West Africa in Precolonial Times," *History in Africa* 19 (1992): 19; Pieter de Marees, *Description and Historical Account of the Gold Kingdom of Guinea (1602)*, trans. and ed. Albert van Dantzig and Adam Jones (Oxford: Oxford University Press, 1987), 162; J. M. F. Nogueira, P. J. P. Fernandes, and A. M. D. Nascimento, "Composition of Volatiles of Banana Cultivars from Madeira Island," *Phytochemical Analysis* 14 (2003): 87–90; F. C. Hoehne, *Botánica e agricultura no Brasil no século XVI* (São Paulo: Companhia Editora Nacional, 1937), 121, 152, 167.

52. S. Douglas Jackson, ed., *The Principal Navigations, Voyages, Traffiques and Dis-*

coveries of the English Nation by Richard Hakluyt, 8 vols. (London: J. M. Dent, 1907), 4:26.

53. The "banana" introduced to the Caribbean by Father Berlanga was presumably known as Dominicos, but this is a common name for a type of plantain, whose presence in the West Indies is independently corroborated for the seventeenth century. Nogueira et al., "Composition of Volatiles," 968.

54. Quoted in Robert Langdon, "The Banana as a Key to Early American and Polynesian History," *Journal of Pacific History* 28 (1993): 21.

55. Crone, *Voyages of Cadamosto,* 70.

56. In the early period of Portuguese overseas voyages, Lisbon occasionally imported rice from West Africa. The metropole recorded deliveries of African rice in 1498, 1506, 1510, and 1514. Vitorino Magalhães Godinho, *Os descobrimentos e a economia mundial,* 2 vols. (Lisbon: Editora Arcádia, 1965) 2:390–93.

57. Viotti da Costa, "Portuguese-African Slave Trade"; A. C. de C. M. Saunders, *A Social History of Black Slaves and Freedmen in Portugal, 1441–1555* (Cambridge: Cambridge University Press, 1982); Vieira, "Sugar Islands," 57–58.

3. AFRICAN FOOD CROPS AND THE GUINEA TRADE

Epigraphs: Alfred W. Crosby, *The Columbian Exchange: Biological and Cultural Consequences of 1492* (Westport, CT: Greenwood Press, 1972), 185; James F. Searing, *West African Slavery and Atlantic Commerce: The Senegal River Valley, 1700–1860* (Cambridge: Cambridge University Press, 1993), 46.

1. Gerald Roe Crone, *The Voyages of Cadamosto and Other Documents on Western Africa in the Second Half of the Fifteenth Century* (London: Hakluyt Society, 1937), 42.

2. Linda M. Heywood and John K. Thornton, *Central Africans, Atlantic Creoles, and the Foundation of the Americas, 1585–1660* (New York: Cambridge University Press, 2007), 60–61.

3. Plow agriculture exposes soils of low fertility to rainfall, thereby accelerating the loss of valuable soil nutrients. Cattle do not thrive in rainforest regions such as that found along the Congo River estuary. "No till" agriculture, performed with a digging stick and handheld hoes, remains the time-honored sustainable farming technique in equatorial environments.

4. As the demand for slaves and food outstripped that for melegueta pepper, and following the discovery of the more pungent chili and black peppers of the Americas and Asia, the terms "Grain Coast" and "Rice Coast" were used interchangeably to indicate the presence of rice. Mpinda, the Atlantic slave port of the Kongo kingdom, was located south of the Congo River estuary; *mpinda* also referred to the indigenous African groundnut *(Vigna subterranea)*. The Bantu term was similarly extended to the New World peanut, after its introduction to the region in the late sixteenth century. From this linguistic origin, the peanut was variously known as pindar in the U.S. South. Andrew Smith, *Peanuts: The Illustrious History of the Goober Pea* (Urbana: University of Illinois Press, 2002), 9.

5. Tambi Eyongetah and Robert Brain, *A History of the Cameroon* (London: Longmans, 1974), 53; P. E. H. Hair, Adam Jones, and Robin Law, *Barbot on Guinea: The Writings of Jean Barbot on West Africa, 1678–1712,* 2 vols. (London: Hakluyt, 1992), 1:319.

6. Mentioned in Linda A. Newson and Susie Minchin, *From Capture to Sale: The Portuguese Slave Trade to Spanish South America in the Early Seventeenth Century* (Boston: Brill, 2007), 79.

7. John Vogt, *Portuguese Rule on the Gold Coast, 1469–1682* (Athens: University of Georgia Press, 1979), 50, 71, 86, 155, 182; Valentim Fernandes, *Description de la côte occidentale d'Afrique* (Bissau, Guinea Bissau: Centro de Estudos da Guiné Portuguêsa, 1951), 190.

8. Pieter de Marees, *Description and Historical Account of the Gold Kingdom of Guinea (1602),* trans. and ed. Albert van Dantzig and Adam Jones (Oxford: Oxford University Press, 1987), pl. 4, 62–64 (plate and text); Adam Jones and P. E. H. Hair, "Sources on Early Sierra Leone: (11) Brun, 1624," *Africana Research Bulletin* 7, no. 3 (1977): 52–64; Hair et al., *Barbot on Guinea,* 2:547; A. M. G. Rutten, *Dutch Transatlantic Medicine Trade in the Eighteenth Century under the Cover of the West India Company* (Rotterdam: Erasmus, 2000).

9. J. D. Fage, "A New Check List of the Forts and Castles of Ghana," *Transactions of the Historical Society of Ghana* 4, no. 1 (1959): 57–67.

10. Ray A. Kea, *Settlements, Trade, and Polities in the Seventeenth-Century Gold Coast* (Baltimore: Johns Hopkins University Press, 1982), 20, 38, 48–50, 57, 301; Georg Nørregård, *Danish Settlements in West Africa, 1658–1850* (Boston: Boston University Press, 1966); K. G. Davies, *The Royal African Company* (New York: Atheneum, 1970 [1957]), 279.

11. Hair et al., *Barbot on Guinea,* 1:170–71; Robert Harms, *The Diligent: A Voyage through the Worlds of the Slave Trade* (New Haven: Yale University Press, 2002), 82, 279; Leif Svalesen, *The Slave Ship Fredensborg* (Bloomington: Indiana University Press, 2000), 99; Theophilus Conneau, *A Slaver's Log Book or 20 Years' Residence in Africa* (Englewood Cliffs, NJ: Prentice-Hall, 1976), 64–65.

12. Heywood and Thornton, *Central Africans,* 186.

13. Searing, *West African Slavery,* 67–68.

14. Bruce Mouser, *A Slaving Voyage to Africa and Jamaica: The Log of the Sandown, 1793–1794* (Bloomington: Indiana University Press, 2002), 56.

15. Rutten, *Dutch Transatlantic Medicine Trade,* 89–90.

16. Eric R. Wolf, *Europe and the People without History* (Berkeley: University of California Press, 1982), 209; Kea, *Settlements,* 12, 48–50, 301; Marees, *Description and Historical Account of the Gold Kingdom of Guinea (1602),* 116, 118; Herbert S. Klein, "The Atlantic Slave Trade to 1650," in Schwartz, *Tropical Babylons,* 223.

17. A majority of the captives sent to the Americas originated in this region. Newson and Minchin, *From Capture to Sale,* 66–67, 106.

18. In the mid-1600s the Jesuit religious order held perhaps as many as ten thousand

slaves, who worked plantations that fed Luanda in Portuguese Angola. Heywood and Thornton, *Central Africans,* 186; Luiz Felipe de Alencastro, *O trato dos viventes: Formação do Brasil no Atlântico Sul; Séculos XVI e XVII* (São Paulo: Companhia das Letras, 2000), 91–95, 251–56; Carlos Sempat Assadourian, *El tráfico de esclavos en Córdoba de Angola a Potosí, siglos XVI–XVII* (Córdoba: Universidade Nacional de Córdoba, 1966), 14.

19. M.D.W. Jeffreys, "How Ancient Is West African Maize?" *Africa* 33, no. 2 (1963): 115–31.

20. Ann Brower Stahl, "Entangled Lives: The Archaeology of Daily Life in the Gold Coast Hinterlands, AD 1400–1900," in Akinwumi Ogundiran and Toyin Falola, eds., *Archaeology of Atlantic Africa and the African Diaspora* (Bloomington: Indiana University Press, 2007), 49–76; Stanley B. Alpern, "The European Introduction of Crops into West Africa in Precolonial Times," *History in Africa* 19 (1992): 13–43.

21. Stahl, "Entangled Lives," 64.

22. The kings of Dahomey along the Slave Coast reserved millet for royal consumption, apparently showing a subsistence preference for the indigenous cereal. See James C. McCann, *Maize and Grace: Africa's Encounter with a New World Crop, 1500–2000* (Cambridge, MA: Harvard University Press, 2005), 33; Adam Jones, *German Sources for West African History, 1599–1669* (Wiesbaden: Verlag, 1983), 222; Kea, *Settlements,* 47–50.

23. Kea, *Settlements,* 301–2.

24. James Barbot, quoted in Hair et al., *Barbot on Guinea,* 2:703n9; see also David Northrup, *Trade without Rulers: Pre-Colonial Economic Development in South-Eastern Nigeria* (Oxford: Oxford University Press, 1978), 171–82.

25. Kea, *Settlements,* 45; Philip D. Curtin, *Economic Change in Pre-Colonial Africa* (Madison: University of Wisconsin Press, 1975), 170. Miracle estimated that the maize demand of the eighteenth-century slave trade may have reached nine thousand tons. Marvin Miracle, *Maize in Tropical Africa* (Madison: University of Wisconsin Press, 1966), 91–92.

26. Dominique Juhé-Beaulaton, "La diffusion du maïs sur les Côtes de l'Or et des esclaves aux XVII et XVIII siècles," *Revue Française d'Historie d'Outre-mer* 77, no. 287 (1990): 177–98; McCann, *Maize and Grace,* 23–24.

27. Attributed to Willem Bosman, in Thomas Astley, *A New General Collection of Voyages and Travels: Consisting of the Most Esteemed Relations, which Have Been hitherto Published in Any Language,* 4 vols. (London: Cass, 1968), 3:58.

28. Juhé-Beaulaton, "Diffusion du maïs." The "millet" described as used for animal feed generally refers to sorghum. Mention of birds as a pest of ripening grain, or the labor spent milling and winnowing, refer to either one of the African cereals, which do not share maize's protective husk. Descriptions of a small-grained cereal that resembles a bulrush are descriptions of pearl millet. Those that report a reddish tinge to the ripened seed head usually refer to sorghum. On the Caribbean plantations where sorghum became an important slave foodstaple, it was called guinea corn.

29. Juhé-Beaulaton, "Diffusion du maïs," 179–80, 190; National Research Council (NRC), *Lost Crops of Africa*, vol. 1, *Grains* (Washington, D.C.: National Academy Press, 1996); CIRAD, *Mémento de l'agronome* (Montpellier, France: CIRAD/Ministère des Affaires Étrangères, 2002).

30. Searing, *West African Slavery*, 140.

31. This was also the case in military conflicts at the turn of the seventeenth century that netted slaves in the kingdom of Kongo. Heywood and Thornton, *Central Africans*, 96.

32. M. Saugnier, quoted in Searing, *West African Slavery*, 124; see also 41, 53, 96–97, 100, 117, 121–22.

33. Hair et al., *Barbot on Guinea*, 1:122–23.

34. Luis da Camara Cascudo, *História da alimentação no Brasil*, 2 vols. (São Paulo: Editora Itatiaia, 1983), 1:209.

35. Claire Robertson and Martin A. Klein, introduction to Robertson and Klein, *Women and Slavery in Africa*, 9. Female slaves also were involved in farming and the spinning and dyeing of cotton cloth that was also traded. Searing, *West African Slavery*, 53, 96–97, 100, 103–4, 117, 122; Boubacar Barry, *Senegambia and the Slave Trade* (Cambridge: Cambridge University Press, 1998), 116–17.

36. Cadamosto, mentioned in Paul Lovejoy, *Transformations in Slavery* (Cambridge: Cambridge University Press, 1983), 32.

37. Kea, *Settlements*, 326; Searing, *West African Slavery*, 46–47.

38. Kate F. Marsters, *Travels in the Interior Districts of Africa by Mungo Park* (Durham, NC: Duke University Press, 2000), 81–82.

39. Barry, *Senegambia*, 107–8, 113, 116; Walter Rodney, "African Slavery and other Forms of Social Oppression on the Upper Guinea Coast in the Context of the Atlantic Slave Trade," in J. E. Inikori, ed., *Forced Migration* (London: African Publishing Company, 1982), 6–70; Martin A. Klein, "Women and Slavery in the Western Sudan," in Robertson and Klein, *Women and Slavery in Africa*, 67–88, esp. 68; Francis Moore, *Travels into the Inland Parts of Africa* (London: Edward Cave, 1738), 43.

40. Gaspard Mollien, *Travels in Africa* (London: Sir Richard Phillips and Co., 1820), 110.

41. Searing, *West African Slavery*, 49, 52–57.

42. Fernandes, quoted in Searing, *West African Slavery*, 22, see also 49, 47.

43. "Rich men of São Tomé had large groups of slaves ranging from 150 to 300 who had the 'obligation to work for their master every day of the week except Sunday, when they worked to support themselves.'" An anonymous early sixteenth-century Portuguese pilot, mentioned in John Thornton, *Africa and Africans in the Making of the Atlantic World, 1400–1680* (New York: Cambridge University Press, 1992), 169–70.

44. De Almada, cited in Walter Rodney, *A History of the Upper Guinea Coast 1545 to 1800* (New York: Monthly Review Press, 1970), 263.

45. Barry, *Senegambia*, 113; Searing, *West African Slavery*, 57.

4. AFRICAN FOOD AND THE ATLANTIC CROSSING

Second epigraph: Barbot's observations were drawn from two slaving voyages (1678–79 and 1681–82) along the western African coast as far south as the island of Príncipe.

1. Trans-Atlantic Slave Trade Database Project, www.metascholar.org/TASTD -Voyages/index.html; David Eltis, S. Behrendt, D. Richardson, and Herbert S. Klein, eds., *The Trans-Atlantic Slave Trade: A Database on CD-ROM* (Cambridge: Cambridge University Press, 1999).

2. Robert Harms, *The Diligent: A Voyage through the Worlds of the Slave Trade* (New Haven: Yale University Press, 2002), 280.

3. P. E. H. Hair, Adam Jones, and Robin Law, eds., *Barbot on Guinea: The Writings of Jean Barbot on West Africa, 1678–1712,* 2 vols. (London: Hakluyt Society, 1992), 1:282n13, 2:681; George Francis Dow, *Slave Ships and Slaving* (Salem, MA: Marine Research Society, 1927), 45, 73; Boubacar Barry, *Senegambia and the Atlantic Slave Trade* (Cambridge: Cambridge University Press, 1998), 118; Elizabeth Donnan, *Documents Illustrative of the History of the Slave Trade to America,* 4 vols. (Washington, D.C.: Carnegie Institution, 1930–35), 1:393–394, 440, 2:192, 247–69, 279–88, 303–4, 3:61, 158, 293, 373–78, 4:530; Bernard Martin and Mark Spurrell, *The Journal of a Slave Trader (John Newton) 1750–1754* (London: Epworth Press, 1962), 20, 27–49, 78–79.

4. Bruce L. Mouser, "Who and Where Were the Baga? European Perceptions from 1793 to 1821," *History in Africa* 29 (2002): 337–64, esp. 356n42; Bruce Mouser, ed., *A Slaving Voyage to Africa and Jamaica: The Log of the Sandown, 1793–1794* (Bloomington: Indiana University Press, 2002), 45, 86n282, 90n295, 99n317.

5. Luiz Felipe de Alencastro, *O trato dos viventes: Formação do Brasil no Atlântico Sul; Séculos XVI e XVII* (São Paulo: Companhia das Letras, 2000), 252.

6. Martin and Spurrell, *Journal of a Slave Trader,* 38, 44, 74; Marcus Rediker, *The Slave Ship: A Human History* (New York: Viking, 2007), 91, 210; Harms, *Diligent,* 279; Herbert S. Klein, *The Atlantic Slave Trade* (Cambridge: Cambridge University Press, 1999), 94; Hair et al., *Barbot on Guinea,* 2:673–74.

7. Alencastro, *Trato dos viventes,* 252.

8. The French-Italian slave trader operated illegally in response to British-led anti–slave trade enforcement in the 1830s and 1840s. He also used Theodore Canot as an alias. Captain Theophilus Conneau, *A Slaver's Log Book or 20 Years' Residence in Africa* (Englewood Cliffs, NJ: Prentice-Hall, 1976), 63–65, 99–103.

9. Arnold R. Highfield and Vladimir Barac, *C. G. A. Oldendorp's History of the Mission of the Evangelical Brethren on the Caribbean Islands of St. Thomas, St. Croix, and St. John* (Ann Arbor, MI: Karoma Publishers, 1987), 214; William Littleton, quoted in James F. Searing, *West African Slavery and Atlantic Commerce: The Senegal River Valley, 1700–1860* (Cambridge: Cambridge University Press, 1993), 140; Philip D. Curtin, *Economic Change in Pre-Colonial Africa* (Madison: University of Wisconsin Press, 1975), 230.

10. Harms, *Diligent,* 279, 281.

11. John Grazilhier to James Barbot, in Hair et al., *Barbot on Guinea*, 2:699–700.

12. Alencastro, *Trato dos viventes*, 255–56.

13. Alexander Falconbridge, *An Account of the Slave Trade on the Coast of Africa* (London: J. Phillips, 1788), 21–22; Conneau, *Slaver's Log Book*, 82; Herbert S. Klein, "The Atlantic Slave Trade to 1650," in Schwartz, *Tropical Babylons*, 201–36, esp. 220; Searing, *West African Slavery*, 140–41; K. G. Davies, *The Royal African Company* (New York: Atheneum, 1970 [1957]), 228.

14. Hair et al., *Barbot on Guinea*, 2:339, 708; Randy Sparks, "The Emergence of Afro-Atlantic Foodways" (paper presented at Cuisines of the Lowcountry and the Caribbean conference, Charleston, SC, March 20–23, 2003); Leif Svalesen, *The Slave Ship Fredensborg* (Bloomington: Indiana University Press, 2000), 111, 119. Fava beans took their common name from their usage in Europe as horse feed.

15. One Portuguese doctor described the food on an eighteenth-century slave ship from Angola to Brazil thus: "the food which is prepared for them is disagreeable and bad-tasting, since they lack the necessary spices, including salt . . . their food consists of nothing more but Indian corn, beans, and manioc flour, all badly prepared and only half cooked." Luiz Antonio de Oliveira Mendes, "A Portuguese Doctor Describes the Suffering of Black Slaves in Africa on the Atlantic Voyage (1793)," in Robert Edgar Conrad, *The Children of God's Fire: A Documentary History of Black Slavery in Brazil* (University Park: Pennsylvania State University Press, 1984), 15–23. Guinea squash and African ackee also formed minor components of some meals. Linda A. Newson and Susie Minchin, *From Capture to Sale: The Portuguese Slave Trade to Spanish South America in the Early Seventeenth Century* (Boston: Brill, 2007), 13; Joseph C. Miller, *Way of Death: Merchant Capitalism in the Angolan Slave Trade, 1730–1830* (Madison: University of Wisconsin Press, 1999), 413–17.

16. For instance, a journey to Sierra Leone from Liverpool typically took six weeks. Rediker, *Slave Ship*, 175.

17. Mahommah Gardo Baquaqua, enslaved in Burkina Faso and shipped to Brazil in the 1840s, wrote that he "suffered very much for lack of water . . . a pint a day was all that was allowed, and no more; and a great many slaves died upon the passage." Mahommah Gardo Baquaqua, "A Young Black Man Tells of His Enslavement in Africa and Shipment to Brazil about the Middle of the Nineteenth Century," in Conrad, *Children of God's Fire*, 27; Harms, *Diligent*, 308–10; Johannes Postma, *The Dutch in the Atlantic Slave Trade, 1600–1815* (Cambridge: Cambridge University Press, 1990), 235.

18. Guinea worm, transmitted by contact with water-borne filaria, was widespread along the Slave Coast. Harms, *Diligent*, 279, 309.

19. Hair et al., *Barbot on Guinea*, 743.

20. Postma, *Dutch in the Atlantic Slave Trade*, 235.

21. The first quote is attributed to Dominican priest Jean-Baptiste Labat, who lived on a plantation in Martinique from the end of the seventeenth century and much of his adult life. Quoted in Thomas Astley, *A New General Collection of Voyages and Travels: Consisting of the Most Esteemed Relations, which Have Been hitherto Published in Any*

Language, 4 vols. (London: Thomas Astley, 1745–47), 3:332. The second quote is from Hair et al., *Barbot on Guinea,* 2:743. On additives to freshen stored water, in general, see A. M. G. Rutten, *Dutch Transatlantic Medicine Trade in the Eighteenth Century under the Cover of the West India Company* (Rotterdam: Erasmus, 2000), 71, 90, 120. On brandy as a water purifier, see Mouser, *Slaving Voyage,* 10.

22. Des Marchais, quoted in Astley, *New General Collection,* 3:57.

23. Jobson, quoted in Astley, *New General Collection,* 3:307; Müller, quoted in Adam Jones, *German Sources for West African History, 1599–1669* (Wiesbaden: Verlag, 1983), 229–30; Palisot-Beauvois (ca. 1805), in Edmund Abaka, "Kola Nut," in Kiple and Ornelas, *Cambridge World History of Food,* 1:684–92, esp. 688.

24. Bosman, quoted in Astley, *New General Collection,* 3:57.

25. Hair et al., *Barbot on Guinea,* 1:188.

26. Abaka, "Kola Nut," 684–92. The kola nut would join the coca leaf in the making of Coca-Cola. Perhaps no other concoction better celebrates the marriage of the African and Amerindian ethnobotanical heritage of the Americas.

27. Newson and Minchin, *From Capture to Sale,* 160.

28. José Flávio Pessoa de Barros and Eduardo Napoleão, *Ewé òrìsà: Uso litúrgico e terapêutico dos vegetais nas casas de candomblé jêje-nago* (Rio de Janeiro: Betrand Brasil, 1998), 282–83; Edward Ayensu, *Medicinal Plants of the West Indies* (Algonac, MI: Reference Publications, 1981), 180; F. G. Cassidy and R. B. Le Page, *Dictionary of Jamaican English* (Kingston, Jamaica: University of the West Indies Press, 2002), 44; H. M. Burkill, *The Useful Plants of West Tropical Africa,* 6 vols. (Kew, England: Royal Botanic Gardens, 1985–2004), 5:92, 617, 653; Lorenzo Dow Turner, *Africanisms in the Gullah Dialect* (Columbia: University of South Carolina Press, 2002), 65.

29. This is because males were presumably better suited to field labor on Caribbean plantations; however, females formed a significant percentage of sugarcane plantation workers. Harms, *Diligent,* 247; and see David Barry Gaspar and Darlene Clark Hine, eds., *More Than Chattel: Black Women and Slavery in the Americas* (Bloomington: Indiana University Press, 1996).

30. Claire Robertson and Martin A. Klein, introduction to Robertson and Klein, *Women and Slavery in Africa,* 9; Searing, *West African Slavery,* 53–54, 96–97, 100, 103–4, 117, 122; Barry, *Senegambia,* 116–17. But also see the discussion of higher female exports from some decentralized West African societies in Martin A. Klein, "The Slave Trade and Decentralized Societies," *Journal of African History* 42 (2001): 49–65; and Walter Hawthorne, *Planting Rice and Harvesting Slaves* (Portsmouth, NH: Heinemann, 2003), 10–14.

31. John Tozer of the *Postillon* at Gambia to the Royal African Company (RAC), May 2, 1704, and Thomas Weaver of the *Swan* to the RAC, May 5, 1704, both quoted in David Eltis, *The Rise of African Slavery in the Americas* (Cambridge: Cambridge University Press, 2000), 106, 106n58.

32. In the early modern period, the English generically referred to most cereals as "corn." Since the slave ship left from Senegal, the grain reference is likely to one of the

African millets, not maize. The first quote is from Hugh Thomas, *The Slave Trade: The Story of the Atlantic Slave Trade; 1440–1870* (New York: Simon and Schuster, 1999), 417; the second quote is from Donnan, *Documents Illustrative,* 3:121, 363–76.

33. The location of the slave stove was often referred to as the airing deck, where the ship's captives were brought up for forced exercise in small groups. Dow, *Slave Ships,* xxiii, 6; Svalesen, *Fredensborg,* 93.

34. Mouser, "Who and Where," 8n32; George Howe, "Last Slave Ship," *Scribner's Magazine* 8, no. 1 (July 1890): 113–29; Dow, *Slave Ships,* xxii.

35. Jay Coughtry, *The Notorious Triangle: Rhode Island and the African Slave Trade, 1700–1807* (Philadelphia: Temple University Press, 1981), 152. The ratio of crew to slaves on these floating prisons was typically higher on American slavers, at one to twelve. See also Rediker, *Slave Ship,* 225.

36. Fonio is also known as *funde.* Newson and Minchin, *From Capture to Sale,* 81–82, 320.

37. Mouser, "Who and Where," 337–64, esp. 356n42; Mouser, *Slaving Voyage,* 45n170, 86n282, 90n295, 99n317; Svalesen, *Fredensborg,* 107.

38. Judith A. Carney, *Black Rice: The African Origins of Rice Cultivation in the Americas* (Cambridge, MA: Harvard University Press, 2001), 25–27, 49–52.

39. Richard Jobson, *The Golden Trade* (Devonshire, England: Speight and Walpole, 1904 [1623]), 68.

40. Carney, *Black Rice,* 108–15.

41. Svalesen, *Fredensborg,* 112; Dow, *Slave Ships,* 66.

42. Pinckard, quoted in Dow, *Slave Ships,* xxiii–xxiv.

43. Smeathman to Drury, Sierra Leone, July 10, 1773, MS D.26 Uppsala University, Sweden, via Starr Douglas, PhD student, Royal Holloway University, London, pers. comm., July 2, 2002.

44. Svalesen, *Fredensborg,* 93, 107.

45. Peter Collinson, "Of the Introduction of Rice and Tar in Our Colonies," *Gentleman's Magazine,* May 26, 1766, 278–80, quoted in A. S. Salley, "Introduction of Rice into South Carolina," *Bulletin of the Historical Commission of South Carolina* (Columbia), no. 6 (1919): 14–16.

46. A. Vaillant, "Milieu cultural et classification des varietés de riz des Guyanes français et hollandaise," *Revue Internationale de Botanique Appliquée et d'Agriculture Tropicale,* no. 33 (1948): 522.

47. Oral histories collected by Judith A. Carney and Rosa Acevedo, historian at the Federal University of Bélen, Brazil, 2002.

48. Simão Estácio da Silveira, *Relação sumária das cousas da Maranhão* (São Paulo: Editora Siciliano, 2001 [1624]), 18. In 1755 the Companhia Geral do Grão-Pará e Maranhão was granted the exclusive right to supply African slaves to the eastern Amazon region. Over the next twenty-two years of its operation, the company shipped more than 25,000 slaves directly from West Africa to Maranhão and Pará. Of the total, nearly 18,000—some 70 percent—originated in the Guinea ports of Bissau and Cacheu (now

Guinea-Bissau). One-third of the total—over 8,000—disembarked at Maranhão to carry out the grueling work of transforming wetland landscapes into rice plantations. António Carreira, *As companhias pombalinas de Grão-Pará e Maranhão e Pernambuco e Paraíba* (Lisboa: Editorial Presença, 1983), 78–79, 96–98; Raimundo José de Sousa Gaioso, *Compêndio histórico-político dos princípios da lavoura do Maranhão* (Paris: Rougeron, 1818), 244–45.

49. Oral histories collected by J. Carney, 1997, and J. Carney and Rosa Acevedo, 2002.

5. MAROON SUBSISTENCE STRATEGIES

Epigraphs: Saramaka Maroon quoted in Richard Price, *First-Time: The Historical Vision of an Afro-American People* (Baltimore: Johns Hopkins University Press, 1983), 71; Ribeiro comment is from Judith Carney fieldwork, Minas Gerais, 2005. This chapter follows the scholarly convention of using the uppercase *M* for Maroons who are recognized as members of distinct and continuing societies. The evidence presented in this chapter is drawn principally from historical and ethnographic studies; archaeological research in maroon sites is less informative on phytolith analysis of cultivated plant remains.

1. Barry Higman, *Slave Populations of the British Caribbean, 1807–1834* (Kingston, Jamaica: University of the West Indies, 1995), 386.

2. The maroon settlement (quilombo) of Palmares was located in the state of Alagoas, Northeast Brazil, about two hundred miles southwest of Recife, Pernambuco. It was in existence at the beginning of the seventeenth century and repeatedly drove back campaigns by Dutch and Portuguese armies until its final destruction in 1694. The population of Palmares at midcentury has been estimated at twenty thousand; it was formed of multiple hamlets that also included Amerindian and mixed-race people. Pedro P. Funari, "The Archaeological Study of the African Diaspora in Brazil," in Ogundiran and Falola, *Archaeology of Atlantic Africa*, 355–71; John Hemming, *Red Gold: The Conquest of the Brazilian Indians, 1500–1760* (Cambridge, MA: Harvard University Press, 1978), 355–57, 363, 369.

3. Official recognition of maroon independence, however, usually required that maroons surrender to colonial authorities any new fugitives who sought refuge in their communities. The first peace treaty with the Maroons in Suriname was made in 1684. Silvia W. de Groot, "Maroons of Surinam: Dependence and Independence," *Annals of the New York Academy of Sciences* 292 (1977): 456. David Isaac de Cohen Nassy lived in the Dutch colony and wrote a history of it, later published in English as Jacob R. Marcus and Stanley F. Chyet, eds., *Historical Essay on the Colony of Surinam, 1788* (Cincinnati: American Jewish Archives, 1974), 56–57. The Ndjuka and Saramakas secured their independence in 1762 after a century of armed struggle with the colonists. Richard Price and Sally Price, *Stedman's Surinam: Life in an Eighteenth-Century Slave Society* (Baltimore: Johns Hopkins University Press, 1992), xix. A similar treaty was signed between the English colonial government and the Leeward Maroons

of Jamaica in 1739. Orlando Patterson, "Slavery and Slave Revolts: A Sociohistorical Analysis of the First Maroon War, 1665 to 1740," in R. Price, *Maroon Societies*, 243–92, esp. 272.

4. An estimated 4.5 million enslaved Africans were brought to Brazil between 1600 and 1850, mostly from west-central Africa. But "the slave population [of Brazil] failed to grow naturally even after the traffic ended in 1851." Robert E. Conrad, *Children of God's Fire: A Documentary History of Black Slavery in Brazil* (University Park: Pennsylvania State University Press, 1994), xx. See also José C. Curto and Paul E. Lovejoy, *Enslaving Connections: Changing Cultures of Africa and Brazil during the Era of Slavery* (Amherst, NY: Humanity Books, 2004), 11; Linda M. Heywood and John K. Thornton, *Central Africans, Atlantic Creoles, and the Foundation of the Americas, 1585–1660* (New York: Cambridge University Press, 2007); and E. Bradford Burns, *A History of Brazil* (New York: Columbia University Press, 1970), 60–61.

5. Ira Berlin and Philip D. Morgan, introduction to Berlin and Morgan, *Cultivation and Culture*, 24.

6. To put the emergence of numerous quilombos into context, slaves working as miners suffered uniformly reduced life spans from the back-breaking labor and the frequent use of mercury in the extraction process. Adelmir Fiabani, *Mato, Palhoça, e Pilão: O quilombo, da escravidão às comunidades remanescentes [1532–2004]* (São Paulo: Editora Expressão Popular, 2005), 215; Kris Lane, "Africans and Natives in the Mines of Spanish America," in Matthew Restall, ed., *Beyond Black and Red: African-Native Relations in Colonial Latin America* (Albuquerque: University of New Mexico Press, 2005), 159–84.

7. Many quilombos formed in mineral-rich Minas Gerais during the eighteenth century. Between 1710 and 1798, government records report that more than 160 quilombos were destroyed. Fiabani, *Mato, Palhoça, e Pilão*, 208.

8. Spoken throughout the vast territory that today includes parts of the Central African Republic, the Democratic Republic of the Congo, and Angola, Kikongo differed from the languages of Angola (Kimbundu and Umbundu) as much as do Portuguese and Spanish. Heywood and Thornton, *Central Africans*, 56.

9. Fiabani, *Mato, Palhoça, e Pilão*, 318–25, 329–30.

10. Carmo da Gama, writing from archival sources in 1904, quoted in Ricardo Ferreira Ribeiro, *Florestas anãs do sertão: O cerrado na história de Minas Gerais* (Belo Horizonte, Brazil: Autêntica Editora, 2005), 312–13; on the expansion of economic interests into the region, see 317.

11. Inácio Correia Pamplona (1731–1810) was a wealthy merchant-landowner who undertook several military expeditions against hostile Indians and quilombolas in western Minas Gerais. Pamplona sought to secure and settle this region in exchange for land grants. Kenneth Maxwell, *Conflicts and Conspiracies: Brazil and Portugal, 1750–1808* (New York: Routledge, 2004), 152–53; Carlos Magno Guimarães, "Esclavage, quilombos et archéologie," *Les Dossiers d'Archéologie*, no. 169 (1992): 67.

12. This account is drawn from Laura de Mello e Souza, "Violência e práticas culturais no cotidiano de uma expedição contra quilombolas: Minas Gerais, 1769," in João José Reis e Flavio dos Santos Gomes, eds., *Liberdade por um fio* (São Paulo: Companhia das Letras, 1996), 193–212, quote on 205–6; R. F. Ribeiro, *Florestas anãs,* 291–343; and Funari, "Archaeological Study," 370.

13. Summarized in R. F. Ribeiro, *Florestas anãs,* 313–18; Mello e Souza, "Violência e práticas culturais." On broader quilombo technology use, see Thomas Flory, "Fugitive Slaves and Free Society: The Case of Brazil," *Journal of Negro History* 64, no. 2 (1979): 116–30, esp. 120–21.

14. These paragraphs are based on fieldwork conducted in 2005 with Jacqueline Chase, geographer at Chico State University.

15. The name is a pseudonym, but the quilombo is located near Milho Verde.

16. João recalls that his male ancestor married an Africa-born woman from Angola, who would have been a linguistically related Kimbundu or Umbundu speaker. More than 1.5 million people from the west-central African region supplied almost half of the captives to the Americas, principally Brazil and Cuba, between 1801 and 1867. Susan J. Herlin, "Brazil and the Commercialization of Kongo, 1840–1870," in Curto and Lovejoy, *Enslaving Connections,* 263.

17. Communal landownership is a defining feature of quilombo communities in Brazil. An individual community member cannot remove part of the landholding for sale to an outsider. The group holds the land for the use of future generations. Carolyn Mooney, "Anthropologist Sheds Light on Jungle Communities Founded by Fugitive Slaves," *Chronicle of Higher Education,* May 22, 1998, B2.

18. Rafael Sanzio Araújo dos Anjos, *Territórios das comunidades quilombolas do Brasil* (Brasilia: Mapas Editora & Consultoria, 2005). Those who oppose these land grants dispute whether all of the black communities requesting land under the quilombo designation are in fact authentically Maroon. It is a fact, however, that descendants of slaves in Brazil have been historically unable to gain land titles. The oral histories, racial composition, and remoteness of many of the communities in Minas Gerais support their argument for true Maroon ancestry. On the percentage of African slaves sent to Brazil, see Roquinaldo Ferreira and Flávio dos Santos Gomes, "African Diaspora Studies in/ and Brazil," paper presented at the African Diaspora Studies and the Disciplines Conference, University of Wisconsin-Madison, March 23–26, 2006.

19. Pero de Magalhães de Gandavo mentioned the *banana de São Tomé* in his book on Brazil, written in 1576. See F. C. Hoehne, *Botánica e Agricultura no Brasil no Século XVI* (São Paulo: Companhia Editora Nacional, 1937), 167.

20. *Momordica charantia* is known in Brazilian Portuguese as *melão de São Caetano.* It is used for improving digestion, for constipation, and as a treatment for malaria. On its pervasive appearance on African diaspora medicinal plant lists, see Judith A. Carney, "African Traditional Plant Knowledge in the Circum-Caribbean Region," *Journal of Ethnobiology* 23, no. 2 (2003): 167–85. On the importance of the guinea fowl in Africa-based religions, see Arno Vogel, Marco Antonio da Silva Mello, and José Flávio

Pessoa de Barros, *A galinha d'Angola: Iniciação e identidade na cultura Afro-Brasileira* (Rio de Janeiro: Universidade Federal Fluminense, 1993).

21. They are believed to share a common origin in West Gondwana prior to tectonic separation some 150 million years ago. Alwyn Gentry, "Diversity and Floristic Composition of Lowland Tropical Forest in Africa and South America," in Peter Goldblatt, ed., *Biological Relationships between Africa and South America* (New Haven: Yale University Press, 1993), 507–47, esp. 512.

22. Fugitive slaves planted rice as a foodstaple in the quilombo of Palmares. Rice and African yams were among the subsistence crops that runaways cultivated in the eighteenth-century Mato Grosso quilombo known as Carlotta. Edison Carneiro, *O Quilombo dos Palmares* (São Paulo: Companhia Editora Nacional, 1958), 20–21; Fiabani, *Mato, Palhoça, e Pilão*, 71.

23. Miguel Esquivel and Karl Hammer, "The Cuban Homegarden 'Conuco': A Perspective Environment for Evolution and In Situ Conservation of Plant Genetic Resources," *Genetic Resources and Crop Evolution* 39, no. 1 (1992): 15; Carney, "African Traditional Plant Knowledge"; Kenneth M. Bilby, *True-Born Maroons* (Gainesville: University of Florida Press, 2005), esp. 129–80.

24. Cyclone Covey, trans., *Cabeza de Vaca's Adventures in the Unknown Interior of America* (Albuquerque: University of New Mexico Press, 1983).

25. See also figure 4.2: this handbill of a slave sale occurring just a few decades later underscores that the arriving Africans had already contracted smallpox. "Coromantee" referred to slaves from the Gold Coast. The variolation practice in Boston's 1721 smallpox outbreak reduced the death rate of those inoculated to just 2 percent. Elizabeth A. Fenn, *Pox Americana* (New York: Hill and Wang, 2001), 32–33; Eugenia Herbert, "Smallpox Inoculation in Africa," *Journal of African History* 16 (1975): 539–59; A. M. G. Rutten, *Dutch Transatlantic Medicine Trade in the Eighteenth Century under the Cover of the West India Company* (Rotterdam: Erasmus, 2000), 101.

26. William Ed Grimé, *Ethno-botany of the Black Americans* (Algonac, MI: Reference Publications, 1979), dedication; Price and Price, *Stedman's Surinam*, 300–302.

27. H. M. Burkill, *The Useful Plants of West Tropical Africa*, 6 vols. (Kew, England: Royal Botanic Gardens, 1985–2004), 5:91–93.

28. Several other genera of pantropical distribution used across the African Atlantic for religious and healing purposes include *Bidens* spp., *Vernonia* spp. (Asteraceae) and *Bauhinia* spp. (Leguminosae, subfamily Caesalpiniacea). *Bidens* is called *ewé susu*, and *Vernonia* is called *ewé auro*. Both vernacular names are from the Yoruba language. On the substitution of well-known African medicinals with taxonomically related New World species by Brazilian slaves, see Robert Voeks, *Sacred Leaves of Candomblé* (Austin: University of Texas Press, 1997), 143–45, 174–77.

29. See the *Bauhinia* species substitution in Brazilian candomblé for the related one of West Africa, in Voeks, *Sacred Leaves*, 143–45; José Flávio Pessoa de Barros and Eduardo Napoleão, *Ewé òrìsà: Uso litúrgico e terapêutico dos vegetais nas casas de candomblé jêje-nago* (Rio de Janeiro: Betrand Brasil, 1998).

30. Morton Marks, "Exploring *El Monte:* Ethnobotany and the Afro-Cuban Science of the Concrete," http://ilarioba.tripod.com/scholars/mortonmarks.htm; Lydia Cabrera, *El Monte* (Miami: Ediciones Universal, 1995 [1954]).

31. On the African bitter melon, see Edward S. Ayensu, *Medicinal Plants of the West Indies* (Algonac, MI: Reference Publications, 1981); Edward S. Ayensu, *Medicinal Plants of West Africa* (Algonac, MI: Reference Publications, 1978). On the guinea fowl, see Vogel et al., *A galinha d'angola.*

32. William Frederick Sharp, *Slavery on the Spanish Frontier: The Colombian Chocó, 1680–1810* (Norman: University of Oklahoma Press, 1976), 18, 20–21, 39, 105, 174; Kenneth Andrews, *The Spanish Caribbean: Trade and Plunder 1530–1630* (New Haven: Yale University Press, 1978), 33; John H. Parry, "Plantation and Provision Ground," *Revista de historia de America* 39 (1955): 13; Judith A. Carney, *Black Rice: The African Origins of Rice Cultivation in the Americas* (Cambridge, MA: Harvard University Press, 2001).

33. Phillip Rowe and Franklin E. Rosales, "Bananas and Plantains," in Jules Janick and James N. Moore, *Fruit Breeding*, vol. 1, *Tree and Tropical Fruits* (New York: John Wiley, 1996), 167–211, esp. 169; W.E. Renkema, *Het Curacaose plantagebedrijf in de negentiende eeuw* (Zutphen, The Netherlands: Walburg, 1981), 114, translation supplied by island anthropologist Rose Mary Allen; Orlando Ribeiro, *Aspectos e problemas da expansão portuguésa* (Lisbon: Estudos de Ciencias Políticas e Sociais, Junta de Investigações do Ultramar, 1962), 49, 88, 116.

34. Parry, "Plantation and Provision Ground," 13.

35. The Dutch and English made the colony transfer as a consequence of the Treaty of Breda in 1667. De Groot, "Maroons of Surinam," 456.

36. Price suspects that the maroons were already growing rice before Paánza's escape and that the seeds commemorated in her story represented a new variety. R. Price, *First-Time,* 89, 129–34.

37. Ibid., 129–34.

38. Most of the enslaved on Suriname's plantations in this period were born in Africa, which remained the case well into the eighteenth century, as slave mortality rates on the colony's sugar plantations remained extremely high. Between 1688 and 1712 some twenty thousand Africans were landed in the colony to work more than 125 plantations. Price and Price, *Stedman's Surinam,* xii; Johannes Postma, *The Dutch in the Atlantic Slave Trade, 1600–1815* (Cambridge: Cambridge University Press, 1990), 183.

39. The seventeenth century was a period of religious upheavals in Europe, which forced the international migration of persecuted minorities such as Jews and certain breakaway sects of Protestants. The Guianas attracted settlement because religious tolerance was upheld on the underpopulated margins of empire. R. Price, *First-Time,* 70–71; Nathalie Zemon Davis, *Women on the Margins: Three Seventeenth-Century Lives* (Cambridge, MA: Harvard University Press, 1995), 172–75; M. Arbell, *The Jewish Nation of the Caribbean* (Jerusalem: Gefen Publishing House, 2002), 29–30, 83, 91.

40. R. Price, *First-Time,* 71.

41. A. Vaillant, "Milieu cultural et classification des varietés de riz des Guyanes

française et hollandaise," *Revue Internationale de Botanique Appliquée et d'Agriculture Tropicale*, no. 33 (1948): 520–29; R. Price, *First-Time*, 129; J. Carney, "'With Grains in Her Hair': Rice History and Memory in Colonial Brazil," *Slavery and Abolition* 25, no. 1 (2004): 1.

42. F. Oudschans Dentz, "De geschiedenis van de rijstbouw in Suriname," *Landbouwkundig Tijdschrift*, no. 691 (1944): 491–92; J. Melville Herskovits and Frances S. Herskovits, *Rebel Destiny: Among the Bush Negroes of Dutch Guiana* (New York: McGraw-Hill, 1934), 93; J. Hurault, *La vie matérielle des noirs réfugiés Boni et des indiens Wayana du Haut-Moroni (Guyane Française): Agriculture économie et habitat* (Paris: ORSTOM, 1965), 27, 38, 92; S. A. Counter and D. L. Evans, *I Sought My Brother: An Afro-American Reunion* (Cambridge, MA: MIT Press, 1981), 167, and images on 168–69; Sally Price, *Co-wives and Calabashes* (Ann Arbor: University of Michigan Press, 1993), 27–32.

43. S. Price, *Co-wives and Calabashes,* 27.

44. Carney, *Black Rice,* 108–13.

45. Vaillant, "Milieu cultural," 522; Roland Portères, "Présence ancienne d'une variété cultivée d'*Oryza glaberrima* en Guyane française," *Journal d'agriculture tropicale et de botanique appliquée* 11, no. 12 (1955): 680. African *glaberrima* rice also appears in collections from former plantation areas of El Salvador and Panama. The British Museum holds another specimen, which was collected in Cuba in 1877, when slavery on the island had not yet ended. Gérard Second, Institut de Recherche pour le Développement, pers. comm., April 19, 2005.

46. R. Price, *First-Time*, 129; Hurault, *Vie matérielle*, 27.

47. J. J. Hartsinck, *Beschrijving van Guiana: Part II* (Amsterdam: S. Emmering, 1974 [1770]), 912. However, yams predominate in the ancestral offerings or memory dishes made by Suriname's Creoles (the population descended from slaves, who were emancipated in 1863). H. Thoden van Velzen (professor of anthropology, University of Utrecht, The Netherlands) and W. van Wetering (research assistant, Free University of Amsterdam, The Netherlands), pers. comm., July 12, 2006.

48. According to Senegal's Serer people, the perforations in vessels symbolize the passage from this world to that of the dead ancestors. Alioune Déme and Ndèye Sokhna Guèye, "Enslavement in the Middle Senegal Valley: Historical and Archaeological Perspectives," in Ogundiran and Falola, *Archaeology of Atlantic Africa*, 122–39. The image of the Kongo burial appears in James H. Sweet, *Recreating Africa: Culture, Kinship, and Religion in the African-Portuguese World, 1441–1770* (Chapel Hill: University of North Carolina Press, 2003), 114; note the presence of chickens, possibly guinea fowl.

49. From "Rebel Village in French Guiana: A Captive's Description," in R. Price, *Maroon Societies,* 317.

50. Richard Price, "Subsistence on the Plantation Periphery: Crops, Cooking, and Labour among Eighteenth-Century Suriname Maroons," *Slavery and Abolition* 12, no. 1 (1991): 107–27.

51. The quote is from Higman, *Slave Populations,* 391. *Tania* or *tannia* interchangeably refers to two tropical tubers, one of New World origin, *Xanthosoma saggi-*

tifolium; the other is Asian taro, *Colocasia esculenta,* which was established as a food crop in Africa centuries, if not millennia, prior to the transatlantic slave trade, whence it was introduced to tropical America.

52. Price and Price, *Stedman's Surinam,* 209–10, 217.

53. Quoted in ibid., 218.

54. Quoted in ibid., 303 (first quote), 216 (second quote).

55. Ibid., 222, 237, 294; John Gabriel Stedman, *Narrative, of a Five Years' Expedition, against the Revolted Negroes of Surinam, in Guiana,* 2 vols. (London: J. Johnson, 1813), 2:108–9, 142, 213, 220–21, 332.

56. Richard Price, introduction to R. Price, *Maroon Societies,* 10; Carney, "African Traditional Plant Knowledge."

6. THE AFRICANIZATION OF PLANTATION FOOD SYSTEMS

1. Just as Europeans in the early modern period referred to any cereal as corn, they called both small- and large-grained African cereals millet. Sorghum grows in areas of higher moisture than the smaller seed plant we today identify as millet. The grain's head is drawn plumed rather than in the form of the cattail or bulrush shape characteristic of millet. The actual drawing appears lifted without attribution from Pieter de Marees' seventeenth-century illustration of the plant (figure 3.5); however, despite Barbot's likely plagiarism, his illustration indicates sorghum was planted as a food crop.

2. Quoted in Robert Voeks, *Sacred Leaves of Candomblé* (Austin: University of Texas Press, 1997), 23.

3. Quentin Buvelot, ed., *Albert Eckhout: A Dutch Artist in Brazil* (The Hague: NIB Capital, 2004); Jorge Marcgrave, *História natural do Brasil* (São Paulo: Imprensa Oficial do Estado, 1942), 43–44, 107–8; F. C. Hoehne, *Botánica e agricultura no Brasil no século XVI* (São Paulo: Companhia Editora Nacional, 1937), 56, 331–32; Frederick Hall, William F. Harrison, and Dorothy Winters Welker, *Dialogues of the Great Things of Brazil* (Albuquerque: University of New Mexico Press, 1987).

4. David Eltis, "Free and Coerced Migrations from the Old World to the New," in David Eltis, ed., *Coerced and Free Migration: Global Perspectives* (Stanford: Stanford University Press, 2002), 33–74, esp. 36; David Eltis, S. Behrendt, D. Richardson, and Herbert S. Klein, eds., *The Trans-Atlantic Slave Trade: A Database on CD-ROM* (Cambridge: Cambridge University Press, 1999); Philip D. Curtin, *The Atlantic Slave Trade* (Madison: University of Wisconsin Press, 1969), 268.

5. The bright red seeds of guinea bean were used as a medicinal, in religious practices, as a unit of weight, in rosaries and necklaces, and as a poison. Hans Sloane, *A Voyage to the Islands Madera, Barbados, Nieves, San Christophers and Jamaica,* 2 vols. (London: Printed by B. M. for the author, 1707–25), 1:181; Arnold R. Highfield and Vladimir Barac, *C. G. A. Oldendorp's History of the Mission of the Evangelical Brethren on the Caribbean Islands of St. Thomas, St. Croix, and St. John* (Ann Arbor, MI: Karoma Publishers, 1987), 120; John Rashford, "Arawak, Spanish and African Contributions to Jamaica's Settlement Vegetation," *Jamaica Journal* 24, no. 3 (1993): 17–23.

6. Voeks, *Sacred Leaves*, 29; João Flavio Pessoa de Barros and Eduardo Napoleão, *Ewé òrìsà: Uso litúrgico e terapêutico dos vegetais nas casas de candomblé jêje-nago* (Rio de Janeiro: Betrand Brasil, 1998), 260.

7. On early Portuguese sources that record "banana" as an African word, see P. K. Reynolds, *The Banana: Its History, Cultivation and Place among Staple Foods* (New York: Houghton Mifflin, 1927), 25–27; H. M. Burkill, *The Useful Plants of West Tropical Africa*, 6 vols. (Kew, England: Royal Botanic Gardens, 1985–2004), 4:225–33.

8. We are indebted to Raul A. Fernandez, professor of social sciences, University of California, Irvine, pers. comm., April 4, 2005, for the reference to Cuba; and to Sarah Fernández, Ecuadorian PhD student at University of California, Los Angeles, pers. comm., May 7, 2007, for that to coastal Ecuador.

9. In seventeenth-century Panama, plantains were regarded as a regular food of Africans. Linda A. Newson and Susie Minchin, *From Capture to Sale: The Portuguese Slave Trade to Spanish South America in the Early Seventeenth Century* (Boston: Brill, 2007), 205.

10. Buvelot, *Albert Eckhout*, 70, 73, 99. The Mexican painting is from an unpublished manuscript by Joaquín Antonio de Basarás y Garaygorta, which dates to 1763.

11. The rise of the banana as the most widely consumed fruit in the West occurred only during the past century, but its commercial success tends to dwarf discussion of the longstanding subsistence role of the plantain in Africa and in the food systems of tropical plantation economies.

12. Karen Ordahl Kupperman, *Providence Island, 1630–1641: The Other Puritan Colony* (Cambridge: Cambridge University Press, 1993), 121.

13. Enslavement of Amerindians was forbidden on the island. Ibid., 166.

14. Ibid., 105–6, 108, 117, 174, 339. The English used "corn" to refer to any cereal, so the reference could include either maize or sorghum. They usually distinguished Amerindian beans from African legumes by calling the latter "peas" after their resemblance to European green peas.

15. In the half century following the arrival of Europeans to the Americas, Amerindian mortality rates reached 90 percent. Charles C. Mann, *1491: New Revelations of the Americas before Columbus* (New York: Knopf, 2005); Judith Carney and Robert Voeks, "Landscape Legacies of the African Diaspora in Brazil," *Progress in Human Geography* 27, no. 2 (2003): 139–52.

16. The first quote is from Jerome S. Handler, "The Amerindian Slave Population of Barbados in the Seventeenth and Early Eighteenth Centuries," *Caribbean Studies* 8, no. 4 (1969): 41; the second quote is from James E. McWilliams, *A Revolution in Eating: How the Quest for Food Shaped America* (New York: Columbia University Press, 2005), 136.

17. Richard Ligon, *A True and Exact History of the Island of Barbadoes* (London: Frank Cass & Co., 1970 [ca. 1647]), 99–100. Taro, or cocoyam, is also known as eddo in parts of the Caribbean.

18. Ligon, *True and Exact History*, 69, 79–82, quote on 46.

19. Karen Ordahl Kupperman, *Roanoke: the Abandoned Colony* (Savage, MD: Rowman and Littlefield, 1984), 17.

20. Lewis C. Gray, *History of Agriculture in the Southern United States to 1860*, 2 vols. (Gloucester, MA: Peter Smith, 1958), 1:16–17.

21. Ligon, *True and Exact History*, 46.

22. McWilliams, *Revolution in Eating*, 133–36; Randy Sparks, "The Emergence of Afro-Atlantic Foodways" (paper presented at Cuisines of the Lowcountry and the Caribbean conference, Charleston, SC, March 20–23, 2003).

23. Luiz Felipe de Alencastro, *O trato dos viventes: Formação do Brasil no Atlântico Sul; Séculos XVI e XVII.* (São Paulo: Companhia das Letras, 2000), 91–94; Carl O. Sauer, *The Spanish Main* (Berkeley: University of California Press, 1966), 53–54, 68–69. Montserrat in the mid-seventeenth century provided the British fleet and army operating in the Leeward Islands with bread made from cassava. Riva Berleant-Schiller and Lydia M. Pulsipher, "Subsistence Cultivation in the Caribbean," *New West Indian Guide* 60, nos. 1–2 (1986): 1–40, esp. 12.

24. The native peoples of the Antilles were for the most part extinct by the 1580s. Alfred W. Crosby, *The Columbian Exchange: Biological and Cultural Consequences of 1492* (Westport, CT: Greenwood Press, 1972), 38; David Watts, *The West Indies: Patterns of Development, Culture and Environmental Change Since 1492* (Cambridge: Cambridge University Press, 1987), 115.

25. Quoted in Kenneth R. Andrews, *The Spanish Caribbean: Trade and Plunder, 1530–1630* (New Haven: Yale University Press, 1978), 33.

26. Paraphrased in John Thornton, *Africa and Africans in the Making of the Atlantic World, 1400–1680* (New York: Cambridge University Press, 1992), 172.

27. J. L. Carstens, quoted in Arnold R. Highfield, *J. L. Carstens' St. Thomas in Early Danish Times* (St. Croix: Virgin Islands Humanities Council, 1997), 72; the second quote is from Waldemar Westergaard, *The Danish West Indies under Company Rule, 1675–1754* (New York: Macmillan, 1917), 158, and see 164–65 on food shortages. See also, James E. McClellan III, *Colonialism and Science: Saint Domingue in the Old Regime* (Baltimore: Johns Hopkins University Press, 1992), 52–53; Dale Tomich, "Une Petite Guinée: Provision Ground and Plantation in Martinique, 1830–1848," in Berlin and Morgan, *Cultivation and Culture*, 224; James H. Sweet, *Recreating Africa: Culture, Kinship, and Religion in the African-Portuguese World, 1441–1770* (Chapel Hill: University of North Carolina Press, 2003), 61.

28. George Warren, *An Impartial Description of Surinam upon the Continent of Guiana in America with a History of Several Strange Beasts, Birds, Fishes, Serpents, Insects and Customs of That Colony, etc.* (London: William Godbid for Nathaniel Brooke, 1667), 19.

29. Joseph C. Miller, "Retention, Reinvention, and Remembering: Restoring Identities through Enslavement in Africa and under Slavery in Brazil," in Curto and Lovejoy, *Enslaving Connections*, 94.

30. Schwartz cites interventions of the Portuguese Crown on behalf of slave sub-

sistence in 1604 and 1701; using primary source documents from the period 1701–11, Conrad shows that the problem continued into the eighteenth century. See Stuart B. Schwartz, "The Mocambo: Slave Resistance in Colonial Bahia," in R. Price, *Maroon Societies*, 206; and Robert Edgar Conrad, *The Children of God's Fire: A Documentary History of Black Slavery in Brazil* (University Park: Pennsylvania State University Press, 1984), 58, 60–61, 161.

31. Robert Harms, *The Diligent: A Voyage through the Worlds of the Slave Trade* (New Haven: Yale University Press, 2002), 357.

32. French official, quoted in Gabriel Debrien, "Marronage in the French Caribbean," in R. Price, *Maroon Societies*, 129; Labat, mentioned in Harms, *Diligent*, 358.

33. Blondel (in 1725), quoted in Harms, *Diligent*, 359.

34. Richard Pares, *Merchants and Planters* (Cambridge: Cambridge University Press, 1960), 39–40. There were other planters who did nothing at all. When drought caused subsistence crops to fail, as occurred in the Danish West Indies from 1725 to 1726, some planters allowed their slaves to starve to death. Waldemar Westergaard, *The Danish West Indies under Company Rule, 1675–1754* (New York: Macmillan, 1917), 165.

35. John H. Parry, "Plantation and Provision Ground," *Revista de historia de América* 39 (1955): 2, 12.

36. William Frederick Sharp, *Slavery on the Spanish Frontier: The Colombian Chocó, 1680–1810* (Norman: University of Oklahoma Press, 1976), 18, 20–21, 39, 105, 174.

37. Hoehne, *Botánica e agricultura*, 121, 152, 167, 186–88, 205–6, 209–11, 218–25.

38. Amerindians revealed the specific properties of autochthonous medicinals, such as Congo root *(Petiveria alliaceae)*. The plant's name recalls not the Amerindians who originally used it, but the enslaved Africans who learned and perpetuated its value. Enslaved women used Congo root to induce abortions. J. Lindley, 1838, mentioned in William Ed Grimé, *Ethno-botany of the Black Americans* (Algonac, MI: Reference Publications, 1979), 156.

39. *Momordica charantia* is the most widely encountered African medicinal in the Americas; it is cultivated for improving digestion, constipation, and as a malaria treatment. Other important species are *Abrus precatorius, Leonotis nepetifolia,* and *Phyllanthus amarus.* Medicinal plants are a very important component of African knowledge systems in the Atlantic world, but are poorly understood from a historical viewpoint. The topic certainly merits separate treatment. Judith Carney, "African Traditional Plant Knowledge in the Circum-Caribbean Region," *Journal of Ethnobiology* 23, no. 2 (2003): 167–85; Edward Ayensu, *Medicinal Plants of the West Indies* (Algonac, MI: Reference Publications, 1981). The prospect for exploring such questions is improving with the recent development of the digital database that contains images and information on all known plants and their uses in Africa and Latin America. The Aluka database (www.aluka.org) is drawn from historical materials found in more than one hundred herbaria in forty countries.

40. Root crops, however, were very important in the Caribbean foodways prior to European contact. Lee A. Newsom and Deborah M. Pearsall, "Trends in Caribbean Is-

land Archaeobotany," in Paul E. Minnis, ed., *People and Plants in Ancient Eastern North America* (Washington, D.C.: Smithsonian Books, 2003), 347–412. On yams, see Parry, "Plantation and Provision Ground," 14.

41. Quoted in Grimé, *Ethno-botany,* 108.

42. The New World yam *(Dioscorea trifida),* of small size, played a far less significant role as food on the island of Jamaica than other domesticates—cassava, sweet potatoes, and arrowroot. The larger African yams were present on Caribbean islands from an early date. John Oldmixon, *The British Empire in America Containing the History of the Discovery, Settlement, Progress and State of the British Colonies of the Continent and Islands of America,* 2 vols. (New York: Augustus M. Kelley, 1969 [1741]), 2:116; F. G. Cassidy and R. B. Le Page, *Dictionary of Jamaican English* (Kingston, Jamaica: University of the West Indies Press, 2002), 328.

43. Kupperman, *Providence Island,* 106; Sloane, *Voyage to the Islands,* 1:lxxix; Woodville K. Marshall, "Provision Ground and Plantation Labor in Four Windward Islands: Competition for Resources during Slavery," in Berlin and Morgan, *Cultivation and Culture,* 203–20; Jerome S. Handler, "Plantation Slave Settlements in Barbados, 1650s to 1834," in Alvin O. Thompson, ed., *In the Shadow of the Plantation* (Kingston, Jamaica: Ian Randle, 2002), 123–161, esp. 161; Clarissa Thérèse Kimber, *Martinique Revisited: The Changing Plant Geographies of a West Indian Island* (College Station: Texas A&M University Press, 1988), 174; David Watts, "Persistence and Change in the Vegetation of Oceanic Islands: An Example from Barbados, West Indies," *Canadian Geographer* 14, no. 1 (1970): 104; Frank W. Pitman, "Slavery on the British West India Plantations," *Journal of Negro History* 11 (1926): 606–8; Newson and Minchin, *From Capture to Sale,* 205.

44. Oviedo's account, in P. K. Reynolds, "Earliest Evidence of Banana Culture," *Journal of the American Oriental Society Supplement,* no. 12, (1951): 1–28, 20–22. In contrast to its colonies, Spain continues to use the word *plátano* to mean the dessert banana, and *plátano de cocinar* for the plantain or cooking banana. Robert Langdon, "The Banana as a Key to Early American and Polynesian History," *Journal of Pacific History* 28 (1993): 15–35; Victor Galán Saúco, Department of Tropical Fruits, Instituto Canario de Investigaciones Agrarias, pers. comm., August 17, 2007.

45. Several researchers have even suggested that the plantain may have arrived in lowland Ecuador via trans-Pacific navigators in pre-Columbian times. Alexander von Humboldt, *Personal Narrative of Travels to the Equinoctial Regions of America, during the Years 1799–1804,* 3 vols. (London: Henry G. Bohn, 1853), 1:189, 191, 2:161, 437; Langdon, "Banana as a Key," 20–24. But see Reynolds, "Earliest Evidence."

46. John Soluri, *Banana Cultures* (Austin: University of Texas Press, 2005), 35–36.

47. On the African origin of the horn plantain, see Edmond de Langhe, R. Swennen, and D. Vuylsteke, "Plantain in the Early Bantu World," *Azania* 29–30 (1994–95): 147–60; R. H. Stover and N. W. Simmonds, *Bananas* (London: Longmans, 1987), 124. On early use of *pacova* in Brazil, see Hoehne, *Botánica e agricultura,* 224; and Hall et al., *Dialogues,* 229n52.

48. Parry, "Plantation and Provision Ground," 39; N. W. Simmonds, *The Evolution of the Bananas* (London: Longmans, 1962), 146; Newson and Minchin, *From Capture to Sale*, 205.

49. Ligon, *True and Exact History*, 32.

50. Ibid., 43, 81.

51. Ibid., 43–44.

52. Ibid., 82.

53. Edmond de Langhe, *Banana and Plantain: The Earliest Fruit Crops?* Annual Report (Montpellier, France: INIBAP, 1995), 8.

54. N. W. Simmonds, *Bananas* (London: Longmans, 1959), 271–72.

55. Ligon, *True and Exact History*, 81.

56. Soluri, *Banana Cultures*, 6, 18–19, 35–36; Robert C. West, *The Pacific Lowlands of Colombia* (Baton Rouge: Louisiana State University Press, 1957), 137.

57. Highfield and Barac, *C. G. A. Oldendorp's History*, 53.

58. Ironmaking, introduced to the Americas, was known in both Africa and Europe prior to the trans-Atlantic slave trade. The quote is from Gray, *History of Agriculture*, 1:195. See also Riva Berleant-Schiller, "The Social and Economic Role of Cattle in Barbuda," *Geographical Review* 67, no. 3 (1977): 299–309; Berleant-Schiller and Pulsipher, "Subsistence Cultivation"; and Watts, *West Indies*, 429.

59. On mainland South America, native peoples were at times able to retreat to remote hinterlands, where over time some managed to develop better resistance to introduced diseases. More than 40 percent of Africans brought to the Americas went to the Caribbean, about the same percentage as to Brazil; but the islands offered far fewer refuges, and the native populations on many of them were exterminated at an early date. Berleant-Schiller and Pulsipher, "Subsistence Cultivation," 19; Eltis et al., *Trans-Atlantic Slave Trade*.

60. In their plant assemblage, the authors pair taro in the same category with New World arrowroot, *Xanthosoma saggitifolium,* as the two are often confused. Berleant-Schiller and Pulsipher, "Subsistence Cultivation," 6, 15, 18.

61. The leaf and its juice are used as a medicine in the Caribbean. Ayensu, *Medicinal Plants of the West Indies,* 140–41; Miguel Esquivel and Karl Hammer, "The Cuban Homegarden 'Conuco': A Perspective Environment for Evolution and In Situ Conservation of Plant Genetic Resources," *Genetic Resources and Crop Evolution* 39, no. 1 (1992): 15; Barros and Napoleão, *Ewé òrìsà,* 149.

7. BOTANICAL GARDENS OF THE DISPOSSESSED

Epigraphs: Guilherme [Willem] Piso, *História natural e médica da India ocidental* (Rio de Janeiro: Instituto Nacional do Livro, 1957 [1645]), 441; Catesby, quoted in William Ed Grimé, *Ethno-botany of the Black Americans* (Algonac, MI: Reference Publications, 1979), 26; Edwards, quoted in Grimé, *Ethno-botany of the Black Americans,* 22.

1. Piso, *História natural e médica,* 441, 443.

2. Jorge Marcgrave, *História natural do Brasil* (São Paulo: Imprensa Oficial do Estado, 1942), 33.

3. The first quote is from Hans Sloane, *A Voyage to the Islands Madera, Barbados, Nieves, San Christophers and Jamaica,* 2 vols. (London: Printed by B. M. for the author, 1707–25), 1:176; the second quote is from Lyman Carrier, *The Beginnings of Agriculture in America* (New York: McGraw-Hill, 1923), 247.

4. Cowpeas have been found in east-central Alabama in a 1700s Creek archaeological site. A legume matching its description (brown, with a black ring around the eye) was known to Le Page du Pratz in Louisiana (1718–34) as the "Apalachean bean," after the Indian group from which it had been obtained. Cowpeas have not been reported archaeologically from Spanish colonial sites in Florida and coastal Georgia or in historical documents of the period. This supports an explanation for their arrival on slave ships but also for African slaves, rather than Spanish colonists, first bringing cowpeas into Native American exchange networks. Kristen Gremillion, "Adoption of Old World Crops and Processes of Cultural Change in the Historic Southeast," *Southeastern Archaeology* 12, no. 1 (1993): 15–20; Elizabeth J. Reitz and C. Margaret Scarry, *Reconstructing Historic Subsistence with an Example from Sixteenth-Century Spanish Florida,* Special Publication Series, no. 3. (Glassboro, NJ: Society for Historical Archaeology, 1985); Carrier, *Beginnings of Agriculture,* 250–51; Robert L. Hall, "Savoring Africa in the New World," in H. J. Viola and C. Margolis, eds., *Seeds of Change* (Washington, D.C.: Smithsonian Institution Press, 1991), 160–71.

5. Catesby was based in Virginia from 1712 to 1719. Quoted in Carrier, *Beginnings of Agriculture,* 246.

6. John Oldmixon, *The British Empire in America Containing the History of the Discovery, Settlement, Progress and State of the British Colonies of the Continent and Islands of America,* 2 vols. (New York: Augustus M. Kelley, 1969 [1741]), 2:116.

7. Antonio Pace, ed., *Luigi Castiglioni's "Viaggio: Travels in the United States of North America, 1785–1787"* (Syracuse: Syracuse University Press, 1983), 171–72.

8. Quoted in Grimé, *Ethno-botany of the Black Americans,* 63.

9. Thomas Jefferson to Anne Cary Randolf, March 22, 1808, in Edwin Morris Betts, ed., *Thomas Jefferson's Garden Book, 1766–1824* (Philadelphia: American Philosophical Society, 1944), 368.

10. Carrier, *Beginnings of Agriculture,* 251.

11. Quoted in Grimé, *Ethno-botany of the Black Americans,* 22.

12. Sloane, *Voyage to the Islands,* 1:184.

13. Ibid., 103 and 105 (first quote), 1:176 (second quote).

14. Quoted in Carrier, *Beginnings of Agriculture,* 246.

15. Thomas Jefferson to Anne Cary Randolf, March 22, 1808, in Betts, *Thomas Jefferson's Garden Book,* 368.

16. Piso, *História natural e médica,* 441.

17. The first quote is from Jasper Danckaerts, quoted in John Thornton, *Africa and Africans in the Making of the Atlantic World, 1400–1680* (New York: Cambridge Uni-

versity Press, 1992), 169; the second quote is also in Thornton, paraphrasing Danckaerts, 169.

18. Quoted in Philip D. Morgan, "Work and Culture: The Task System and the World of Low Country Blacks, 1700 to 1880," *William and Mary Quarterly*, 3rd Series, vol. 39, no. 4 (1982): 565.

19. Sloane, *Voyage to the Islands*, 1:103; Mark W. Hauser, "Between Urban and Rural: Organization and Distribution of Local Pottery in Eighteenth-Century Jamaica," in Ogundiran and Falola, *Archaeology of Atlantic Africa*, 300.

20. Luis da Camara Cascudo, *História da alimentação no Brasil*, 2 vols. (São Paulo: Editora Itatiaia, 1983), 1:228; F. C. Hoehne, *Botánica e agricultura no Brasil no século XVI* (São Paulo: Companhia Editora Nacional, 1937), 56, 331–32; Soares, mentioned in Cascudo, *História da alimentação*, 1:228.

21. Moreau, paraphrased in Thornton, *Africa and Africans*, 174.

22. Thornton, *Africa and Africans*, 170–74; Ciro Flamarion S. Cardoso, "The Peasant Breach in the Slave System: New Developments in Brazil," *Luso-Brazilian Review* 25, no. 1 (1988): 49–57.

23. James F. Searing, *West African Slavery and Atlantic Commerce: The Senegal River Valley, 1700–1860* (Cambridge: Cambridge University Press, 1993), 22, 57.

24. Robert Edgar Conrad, *The Children of God's Fire: A Documentary History of Black Slavery in Brazil* (University Park: Pennsylvania State University Press, 1984), 58, 60–62, 154, 161.

25. Riva Berleant-Schiller, "Grazing and Gardens in Barbuda," in Riva Berleant-Schiller and Eugenia Shanklin, eds., *The Keeping of Animals: Adaptation and Social Relations in Livestock Producing Communities* (Totowa, NJ: Allanheld and Osmun, 1983), 80.

26. Stuart B. Schwartz, "Resistance and Accommodation in Eighteenth Century Brazil: The Slaves' View of Slavery," *Hispanic American Historical Review* 57, no. 1 (1977): 69–81, quotes on 77 and 79.

27. Quoted in Cardoso, "Peasant Breach," 55

28. Conrad, *Children of God's Fire*, xx.

29. In Jamaica, conditions were even worse for sugar slaves, who typically toiled an average annual 4,000 hours in field labor, which did not include the time spent on tending provision grounds. Higman compares this to the average annual 2,900 hours for a British factory worker around 1830. Barry Higman, *Slave Populations of the British Caribbean, 1807–1834* (Jamaica: University of the West Indies, 1995), 188.

30. Arnold R. Highfield, and Vladimir Barac, *C. G. A. Oldendorp's History of the Mission of the Evangelical Brethren on the Caribbean Islands of St. Thomas, St. Croix, and St. John* (Ann Arbor, MI: Karoma Publishers, 1987), 222.

31. John Michael Vlach, *By the Work of Their Hands: Studies in Afro-American Folklife* (Charlottesville: University Press of Virginia, 1991), 220.

32. Harold C. Syrett, ed., *The Papers of Alexander Hamilton*, 27 vols. (New York: Columbia University Press, 1963), 1:61–62.

33. Lydia M. Pulsipher, "The Landscapes and Ideational Roles of Caribbean Slave Gardens," in N. Miller and K. L. Gleason, eds., *The Archaeology of Garden and Field* (Philadelphia: University of Pennsylvania Press, 1994), 202–22.

34. Cattle ranching was an early feature of the Carolina colony. In 1750 there was an estimated 100,000 head of cattle, many of which were slaughtered and exported as salt beef to the British West Indies. J. S. Otto and Nain E. Anderson, "Cattle Ranching in the Venezuelan Llanos and the Florida Flatwoods: A Problem in Comparative History," *Comparative Studies in Society and History* 28, no. 4 (1986): 672–83, esp. 680–81.

35. The black-eyed pea was the third most important food crop produced on Carolina plantations between 1730 and 1776, after rice and indigo. The African introduction remains an important rotation crop in South Carolina. Laurens shipped his black-eyed peas to St. Kitts and Grenada in the 1750s. Randy Sparks, "The Emergence of Afro-Atlantic Foodways" (paper presented at Cuisines of the Lowcountry and the Caribbean conference, Charleston, SC, March 20–23, 2003); Philip Morgan, *Slave Counterpoint: Black Culture in the Eighteenth-Century Chesapeake and Lowcountry* (Chapel Hill: University of North Carolina Press, 1998), 50.

36. Barbuda was held as an exclusive lease by the English Codrington family from 1685 to 1870, and they used it to supply their Antiguan sugar estates. Berleant-Schiller, "Grazing and Gardens," 79.

37. "Eddoes" interchangeably meant taro, introduced from Africa, as well as the New World edible root crop *Xanthosoma* spp. Sidney W. Mintz, professor emeritus of anthropology, John Hopkins University, pers. comm., June 21, 2007; R. A. J. van Lier, *Frontier Society: A Social Analysis of the History of Surinam* (The Hague: Martinus Nijhoff, 1971), 26–27; A. van Stipriaan, *Surinaams Contrast* (Leiden: KITLV Uitgeverij, 1993), 350–57; Waldemar Westergaard, *The Danish West Indies under Company Rule, 1675–1754* (New York: Macmillan, 1917), 158.

38. Barbara J. Heath and Amber Bennett, "'The little Spots allow'd them': The Archaeological Study of African-American Yards," *Historical Archaeology* 34, no. 2 (2000): 40; Sidney W. Mintz and Douglas Hall, *The Origins of the Jamaican Internal Marketing System,* Yale University Publications in Anthropology 57 (New Haven: Yale University Press, 1960), 10.

39. Ira Berlin and Philip D. Morgan, introduction to Berlin and Morgan, *Cultivation and Culture,* 1–45, esp. 26–27; Higman, *Slave Populations,* 204.

40. Berlin and Morgan, introduction to Berlin and Morgan, *Cultivation and Culture,* 23, 28; John H. Parry, "Plantation and Provision Ground," *Revista de historia de America* 39 (1955): 1–20, esp. 16.

41. Higman, *Slave Populations,* 204, 211; Berlin and Morgan, introduction to Berlin and Morgan, *Cultivation and Culture,* 28.

42. The first quote is from Thornton, *Africa and Africans,* 169–70; the second quote is from Walter Rodney, *A History of the Upper Guinea Coast, 1545 to 1800* (New York: Monthly Review Press, 1970), 236.

43. Searing, *West African Slavery,* 53–57.

44. Searing, *West African Slavery*, 53–57, 117; Boubacar Barry, *Senegambia and the Slave Trade* (Cambridge: Cambridge University Press, 1998), 113.

45. This body of work is known as the "peasant breach" literature. The name refers to the lands that some plantation societies granted slaves for their independent production. Slaves were granted time to work on the small plots and could sell the surplus food they produced. Some scholars have argued that this form of independent production provided an opening (niche or breach) in the master-slave relationship by offering slaves control over a portion of their lives. However, the slave economy never formed an independent logic of its own and remained tied to white hegemony and control. Sidney W. Mintz, "Slavery and the Rise of Peasantries," *Historical Reflections* 6 (Summer 1979): 213–42; Mintz and Hall, *Origins of the Jamaican*; Cardoso, "Peasant Breach"; Dale Tomich, "Une Petite Guinée: Provision Ground and Plantation in Martinique, 1830–1848," in Berlin and Morgan, *Cultivation and Culture*, 221–42, esp. 222.

46. Woodville K. Marshall, "Provision Ground and Plantation Labor in Four Windward Islands: Competition for Resources during Slavery," in Berlin and Morgan, *Cultivation and Culture*, 215–16.

47. Quoted in Conrad, *Children of God's Fire*, 190.

48. Schoelcher's observation that slaves worked their subsistence plots "communally" is quoted in Tomich, "Petite Guinée," 234. Schoelcher also observed individual ownership of trees on Martinique, which is still a common tenure practice in rural West Africa. The person who plants a tree holds the right to its fruit, even if it grows on land held by another, and the tree owner can pass on those rights to a person of his or her choosing. "There are some planters who do not have fruit trees on their plantations because tradition establishes that such and such a tree belongs to such and such a Negro, and they [the planters] have little hope of ever enjoying them because the slave bequeaths his tree just like the rest of his property." Quoted in ibid., 235. However, after emancipation such rights were often abrogated. Pulsipher, for instance, recounts how ex-slaves on Montserrat were heavily fined for picking fallen fruit off the ground. Pulsipher, "Landscapes and Ideational Roles," 215.

49. Jean Besson, "Family Land and Caribbean Society: Toward an Ethnography of Afro-Caribbean Peasantries," in E. Thomas-Hope, ed., *Perspectives on Caribbean Regional Identity* (Liverpool: Centre for Latin American Studies, University of Liverpool, 1984), 57–83.

50. In eighteenth-century Jamaica, these slave provision grounds were known as *polincks* and located in "hills up to ten miles from their houses, which they cleared and tended themselves." Edward Brathwaite, *The Development of Creole Society in Jamaica, 1770–1820* (Oxford: Clarendon, 1971), 133. See also Heath and Bennett, "'Little Spots'"; Mintz and Hall, *Origins of the Jamaican*; Pulsipher, "Landscapes and Ideational Roles," 218.

51. Pinckard, quoted in Jerome S. Handler, "Plantation Slave Settlements in Barbados, 1650s to 1834," in Alvin O. Thompson, ed., *In the Shadow of the Plantation* (Kingston, Jamaica: Ian Randle, 2002), 133n4; George Pinckard, *Notes on the West*

Indies (London: Longman, Hurst, Rees, and Orme, 1806); Heath and Bennett, "'Little Spots.'" 40.

52. Berlin and Morgan, introduction to Berlin and Morgan, *Cultivation and Culture*, 29–30; Higman, *Slave Populations*, 222; Richard Ligon, *A True and Exact History of the Island of Barbadoes* (London: Frank Cass & Co., 1970), 24; Marcgrave, *História natural*, 43; Piso, *História natural e médica*, 527; Cascudo, *História da alimentação*, 186, 240; Highfield and Barac, *C. G. A. Oldendorp's History*, 115, 120–21, 660, 671; David Watts, "Persistence and Change in the Vegetation of Oceanic Islands: An Example from Barbados, West Indies," *Canadian Geographer* 14, no. 2 (1970): 91–109; A. M. G. Rutten, *Dutch Transatlantic Medicine Trade in the Eighteenth Century under the Cover of the West India Company* (Rotterdam: Erasmus, 2000), 67, 72, 88, 98–99; Edmund Abaka, "Kola Nut," in Kiple and Ornelas, *Cambridge World History of Food*, 1:684–92; Brathwaite, *Development of Creole Society*, 133.

53. Quoted in Tomich, "Petite Guinée," 222.

8. GUINEA'S PLANTS AND EUROPEAN EMPIRE

Epigraph: A. S. Salley, "Introduction of Rice into South Carolina," *Bulletin of the Historical Commission of South Carolina* (Columbia), no. 6 (1919), 15.

1. While there are also New World and Asian yam species, the reference is to a yellow yam introduced from the African continent. Another white African yam *(D. rotundata)* was also introduced and consumed. The American yam was a minor food-staple among Amerindian populations; the Asian yam *(D. alata)*, known as the greater yam for its immense tubers (sometimes over 100 pounds), was also subsequently introduced to the Caribbean. The summary of Oviedo's account (1526) is from John H. Parry, "Plantation and Provision Ground," *Revista de historia de America* 39 (1955): 13. See also J. Alexander and D. G. Coursey, "The Origins of Yam Cultivation," in Peter Ucko and G. W. Dimbleby, eds., *The Domestication and Exploitation of Plants and Animals* (London: Duckworth, 1969), 405–25.

2. The Mande language refers to the yam as *yambi* or *niam*, Temne as *enyame*. In the Fula language, *nyam* means "to eat" and *nyammi* means "food." Along the Lower Guinea Coast, the word for yam is *nyam* (Twi), *iyá* (Yoruba), and *ashnama* (Kwa). H. I. Burkill, "The Contact of the Portuguese with African Food Plants which Gave Words such as 'Yam' to European Languages," *Proceedings of the Linnean Society of London*, part 2 (1938): 84–95; John T. Schneider, *Dictionary of African Borrowings in Brazilian Portuguese* (Hamburg: Helmut Buske Verlag, 1991); F. G. Cassidy and R. B. Le Page, *Dictionary of Jamaican English* (Kingston, Jamaica: University of the West Indies Press, 2002), 325–26; Roger M. Blench, *Archaeology, Language, and the African Past* (Lanham, MD: Altamira Press, 2006), 224.

3. Redcliffe Salaman, *The History and Social Influence of the Potato* (Cambridge: Cambridge University Press, 1970), 112.

4. H. M. Burkill, *The Useful Plants of West Tropical Africa*, 6 vols. (Kew, England: Royal Botanic Gardens, 1985–2004), 4:918; Cassidy and Le Page, *Dictionary of Ja-*

maican English, 328–29; Douglas B. Chambers, *Murder at Montpelier: Igbo Africans in Virginia* (Jackson: University Press of Mississippi, 2005), 40.

5. Johann Baptist von Spix and Carl Friedrich Philipp von Martius, *Travels in Brazil, in the Years 1817–1820,* 2 vols. (London: Longman, Hurst, Rees, Orme, Brown, and Green, 1824), 2:98; Richard Price, "Subsistence on the Plantation Periphery: Crops, Cooking, and Labour among Eighteenth-Century Suriname Maroons," *Slavery and Abolition* 12, no. 1 (1991): 107–27; J. T. Schneider, *Dictionary of African Borrowings,* 156, 246.

6. Jorge Marcgrave, *História natural do Brasil* (São Paulo: Imprensa Oficial do Estado, 1942), 21.

7. Thomas Jefferson to John Taylor, June 23, 1808, in Edwin Morris Betts, ed., *Thomas Jefferson's Garden Book, 1766–1824* (Philadelphia: American Philosophical Society, 1944), 372.

8. Lorenzo Dow Turner, *Africanisms in the Gullah Dialect* (Columbia: University of South Carolina Press, 2002), 62; Maureen Warner-Lewis, *Guinea's Other Suns: The African Dynamic in Trinidad Culture* (Dover, MA: Majority Press, 1991), 169; Guilherme [Willem] Piso, *História natural e médica da India ocidental* (Rio de Janeiro: Instituto Nacional do Livro, 1957 [1645]), 441; H. M. Burkill, *Useful Plants,* 4:797. Commercial sesame is of Asian origin, but related and wild species of the plant have been cultivated and collected in Africa since ancient times.

9. Mary Tolford Wilson, "Peaceful Integration: The Owner's Adoption of His Slaves' Food," *Journal of Negro History* 49, no. 2 (1964): 116–27, esp. 119.

10. Thomas Jefferson to Anne Cary Randolf, March 22, 1808, in Betts, *Thomas Jefferson's Garden Book,* 368.

11. J. T. Schneider, *Dictionary of African Borrowings,* 157; Cassidy and Le Page, *Dictionary of Jamaican English,* 216. The Dutch Caribbean name appears in Thomas Astley, *A New General Collection of Voyages and Travels: Consisting of the Most Esteemed Relations, which Have Been hitherto Published in Any Language,* 4 vols. (London: Thomas Astley, 1745–47), 3:219.

12. Cassidy and Le Page, *Dictionary of Jamaican English,* 44; H. M. Burkill, *Useful Plants,* 5:92, 617, 653; Turner, *Africanisms,* 65.

13. J. T. Schneider, *Dictionary of African Borrowings,* 129; Cassidy and Le Page, *Dictionary of Jamaican English,* 2.

14. Cassidy and Le Page, *Dictionary of Jamaican English,* 168–69.

15. H. M. Burkill, *Useful Plants,* 4:225–29, 792–93.

16. Marcgrave, *História natural,* 43–44; R. Price, "Subsistence on the Plantation Periphery," 107–27.

17. "The second sort of subterraneous Beans, have been known to us but a few Years, and are called *Angola* Beans, by reason they were transplanted from thence to this place." Willem Bosman, *A New and Accurate Description of the Coast of Guinea, Divided into the Gold, the Slave, and the Ivory Coasts* (London: Printed for James Knapton and Dan. Midwinter, 1705), 301.

18. Marcgrave, *História natural,* 43–44; Luis da Camara Cascudo, *História da ali-*

mentação no Brasil, 2 vols. (São Paulo: Editora Itatiaia, 1983), 1:186–87, 245; National Research Council (NRC), *Lost Crops of Africa,* vol. 2, *Vegetables* (Washington, D.C.: National Academy Press, 2006), 53–72.

19. R. Price, "Subsistence on the Plantation Periphery," 110; Burkill, *Useful Plants,* 3:735. Vernacular names for the Bambara groundnut along the former Gold and Slave coasts include *gub-a-gube, gub-gub,* and *guba-gobgob.* These in turn are likely derivations of the Kikongo-language word *nguba.* Kikongo is a Bantu language spoken in the Congo and Angola regions.

20. Bosman, *New and Accurate Description,* 301.

21. Cassidy and Le Page, *Dictionary of Jamaican,* 351; Turner, *Africanisms,* 149; Andrew Smith, *Peanuts: The Illustrious History of the Goober Pea* (Urbana: University of Illinois, 2002), 3, 9.

22. Hans Sloane, *A Voyage to the Islands Madera, Barbados, Nieves, San Christophers and Jamaica,* 2 vols. (London: Printed by B.M. for the author, 1707–25), 1:184.

23. Henry Barham, *Hortus Americanus* (Kingston, Jamaica: Alexander Aikman, 1794), 145–46.

24. This is St. Catherine Parish, located in the country's southeast. E. Kofi Agorsah, "Scars of Brutality: Archaeology of the Maroons in the Caribbean," in Ogundiran and Falolfa, *Archaeology of Atlantic Africa,* 340–42.

25. Lewis Gray, *History of Agriculture in the Southern United States to 1860,* 2 vols. (Gloucester, MA: Peter Smith, 1958), 2:828; William Watson, "Some Account of an Oil, Transmitted by Mr. George Brownrigg, of North Carolina [October 31, 1769]," in *Philosophical Transactions of the Royal Society of London,* vol. 59, by the Royal Society (Great Britain), Charles Hutton, George Shaw, and Richard Pearson (published and printed by and for C. and R. Baldwin, 1809), 665–67.

26. African American vendors are credited with popularizing the peanut as a snack food in the North during the nineteenth century. A. Smith, *Peanuts,* 14–17.

27. American colonists used animal lard in their cooking, as a consequence of the vast animal herds kept. Europeans, on the other hand, depended considerably more on vegetable oil imported from Mediterranean countries. A. Smith, *Peanuts,* 65.

28. Quoted in W. Watson, "Some Account," 665.

29. The peanut's use in making cooking oil was increasingly recognized following the batch sent to England in 1769, which did not go rancid after more than an eight-month sea voyage. W. Watson, "Some Account."

30. Africans crossed erect and prostrate varieties of the peanut, from which several desirable types, or landraces, developed. From this experimentation, western Africa emerged a secondary center of variation of the peanut. The Virginia peanut cultivar likely developed from the African landraces that accompanied slaves to North America. Jonathon D. Sauer, *Historical Geography of Crop Plants* (Boca Raton, FL: CRC Press, 1993), 82–83; H.M. Burkill, *Useful Plants,* 3:288.

31. The boiled peanut was often made into a dish with guinea squash *(Solanum*

aethiopicum) in British America. John Martin Taylor, "Boiled Peanuts," *Gastronomica* 1, no. 4 (2001): 25–26. Raw peanuts combined with sugar syrup were also made into the hard candy known in English as peanut brittle. In Brazil, peanut brittle is not called by an Amerindian cognate, but rather by a derivative associated with the presence of enslaved Africans: *pé de moleque,* "foot of the street urchin" (the term *moleque* was used derogatorily in slavery to refer to black children). This confection is prepared in Senegambia and across diasporic communities.

32. M. T. Wilson, "Peaceful Integration," 116–17.

33. Ibid.

34. NRC, *Lost Crops of Africa,* vol. 1, *Grains* (Washington, D.C.: National Academy Press, 1996), 79, 127–30, 148.

35. Sorghum remains to this day an important animal feed in the Americas.

36. Charles Bryant, *Flora Diaetetica: Or, History of Esculent Plants, Both Domestic and Foreign; In which They Are Accurately Described, and Reduced to Their Linnaean Generic and Specific Names, with Their English Names Annexed* (London: B. White, 1783), 336.

37. Woodville K. Marshall, "Provision Ground and Plantation Labor in Four Windward Islands: Competition for Resources during Slavery," in Berlin and Morgan, *Cultivation and Culture,* 203–20; David Harris, *Plants, Animals, and Man in the Outer Leeward Islands, West Indies: An Ecological Study of Antigua, Barbuda, and Anguilla,* University of California Publications in Geography 18 (Berkeley: University of California Press, 1965), 92.

38. Frederick Hall, William F. Harrison, and Dorothy Winters Welker, *Dialogues of the Great Things of Brazil* (Albuquerque: University of New Mexico Press, 1987), 225n26.

39. Sloane, *Voyage to the Islands,* 1:104.

40. Ibid., 1:104–5.

41. Hugh Talmage Lefler, ed., *A New Voyage to Carolina by John Lawson* (Chapel Hill: University of North Carolina Press, 1967 [1709]), 82.

42. In some of these locations, slaves raised livestock; in others, they extracted salt or were employed in fishing, capturing turtles, or as pearl divers.

43. The cattle keepers of Barbuda still practice a customary form of land tenure in which the grazing land is open and not held privately. This is also the practice in the traditional Fula (Fulbe) pastoral system of West Africa. Riva Berleant-Schiller, "Grazing and Gardens in Barbuda," in Riva Berleant-Schiller and Shanklin, eds., *The Keeping of Animals: Adaptation and Social Relations in Livestock Producing Communities* (Totowa, NJ: Allanheld and Osmun, 1983), 73–91; D. R. Harris, *Plants, Animals and Man,* 81, 91, 100, 110; Andrew Sluyter, "The Role of Black Barbudans in the Establishment of Open-Range Cattle Herding in the Colonial Caribbean and South Carolina," *Journal of Historical Geography* 35 (2009): 330–49.

44. The particular food delivered was not specified. Company records indicate there were three thousand slaves on Curaçao that year. Johannes Postma, *The Dutch in the*

Atlantic Slave Trade, 1600–1815 (Cambridge: Cambridge University Press, 1990), 33, 43–44; Johannes Postma, "West-African Exports and the Dutch West India Company, 1675–1731," *Economisch- en Sociaal-Historisch Jaarboek* 36 (1973): 65, 73.

45. There is nothing else to indicate whether manumission did in fact occur. M. A. Visman, "Van slaaf tot plantagehouder. Een aspect van het 18e eeuws plantagewezen op Curaçao," *Nieuw West-Indische Gids* 55, nos. 1–2 (1981): 41, translation supplied by island anthropologist Rose Mary Allen; H. Bloom, *The Economic Activities of the Jews of Amsterdam in the Seventeenth and Eighteenth Centuries* (Williamsport, PA: Bayard Press, 1937), 141–42.

46. From the newspaper article, Anon., "Het guineesche graan," *Curaçaosche Courant,* January 20, 1838, translation supplied by Rose Mary Allen. See also Wim Klooster, *Illicit Riches: Dutch Trade in the Caribbean, 1648–1795* (Leiden: KITLV Press, 1998), 105–6; and Han Jordaan, "The Curaçao Slave Market: From *Asiento* Trade to Free Trade, 1700–1730," in Postma and Enthoven, *Riches from Atlantic Commerce,* 219–57, esp. 222, 237–38.

47. Governor Kikker called what was planted "millet," but the crop's six-month cultivation cycle and accompanying description indicate that it was actually sorghum. W. E. Renkema, *Het Curacaose plantagebedrijf in de negentiende eeuw* (Zutphen: Walburg, 1981), 114, translation supplied by Rose Mary Allen.

48. The sorghum harvest formerly coincided with the Easter/Passover period at the end of March/early April. Today, the *seú* festival, as islanders call it, is a carnival. Its original meaning as a harvest festival is now entirely obscured. Oil refineries have contributed to the drop of the island's water table and deepening desiccation, and the cultivation of food crops in practically any form has diminished significantly. Renkema, *Het Curacaose.*

49. Anon., "Het guineesche graan."

50. Harvest festivals were a traditional feature of many West African societies. Lynn Brydon, "Rice, Yams and Chiefs in Avatime: Speculations on the Development of a Social Order," *Africa* 51, no. 2 (1981): 659–77; D. G. Coursey and C. K. Coursey, "The New Yam Festivals of West Africa," *Anthropos* 66, nos. 3–4 (1971): 444–84. Yam harvest festivals are still celebrated in Jamaica, although it is unclear whether this was also the case in slavery times.

51. The account mentions that the crop took five to six months to mature, another indication that it was sorghum, not millet. Arnold R. Highfield and Vladimir Barac, *C. G. A. Oldendorp's History of the Mission of the Evangelical Brethren on the Caribbean Islands of St. Thomas, St. Croix, and St. John* (Ann Arbor, MI: Karoma Publishers, 1987), 107, 132–34.

52. Edward Long, *History of Jamaica,* 3 vols. (London: Lowndes, 1774), 2:762.

53. Barry Higman, *Slave Populations of the British Caribbean, 1807–1834* (Kingston, Jamaica: University of the West Indies, 1995), 204–7.

54. Ibid., 206.

55. Bryant, *Flora Diaetetica,* 336.

56. George Pinckard's visit to Barbados and the hymn are mentioned in Higman, *Slave Populations*, 206–7.

57. The Indian trade shifted quickly in the early colonial period from a trade with Native Americans for deerskin to a slave trade in native peoples. By promoting intertribal hostility, the colonists made money from selling slaves, while depopulating the land and opening it up for the expansion of European settlement. In 1708 there were about 5,300 whites in the settlement, 2,900 African slaves, and 1,400 Indian slaves. Others were deported to plantations of the West Indies. Gary B. Nash, *Red, White and Black: The Peoples of Early North America* (Englewood Cliffs, NJ: Prentice-Hall, 1992), 131–32, 140.

58. James E. McWilliams, *A Revolution in Eating: How the Quest for Food Shaped America* (New York: Columbia University Press, 2005), 133–35.

59. Quoted in Lyman Carrier, *The Beginnings of Agriculture in America* (New York: McGraw-Hill, 1923), 249.

60. The earliest eggplant grown in the Americas was the African eggplant *(Solanum aethiopicum)*. Widely grown as a food crop in the plantation period, it was eventually replaced by its Asian cousin, *S. melongena*. Peter H. Wood, *Black Majority* (New York: Knopf, 1974), 120n92; NRC, *Lost Crops,* 2:137; Grimé, *Ethno-botany of the Black Americans,* 26.

61. Lefler, *New Voyage to Carolina,* 82.

62. Mark Catesby, *The Natural History of Carolina, Florida, and the Bahama Islands,* 2 vols. (London, 1771 [1743]), 1:xviii.

63. Randy Sparks, "The Emergence of Afro-Atlantic Foodways" (paper presented at Cuisines of the Lowcountry and the Caribbean conference, Charleston, SC, March 20–23, 2003).

64. Raymond Breton (in 1665), referenced in Carrier, *Beginnings of Agriculture,* 247–48.

65. NRC, *Lost Crops,* 2:115–16; CIRAD, *Mémento de l'agronome* (Montpellier, France: CIRAD/Ministère des Affaires Étrangères, 2002); Sparks, "Emergence of Afro-Atlantic Foodways."

66. M. T. Wilson, "Peaceful Integration," 120.

67. William Douglass, quoted in Carrier, *Beginnings of Agriculture,* 250.

68. Morgan, referenced in Sparks, "Emergence of Afro-Atlantic Foodways"; Philip D. Morgan, *Slave Counterpoint: Black Culture in the Eighteenth-Century Chesapeake and Lowcountry* (Chapel Hill: University of North Carolina Press, 1998), 50.

69. Leslie Stephen and Sidney Lee, eds., *Dictionary of National Biography* (New York: Macmillan, 1893), 209.

70. Gray, *History of Agriculture,* 1:194.

71. Governor Berkeley, quoted in Daniel C. Littlefield, *Rice and Slaves* (Baton Rouge: Louisiana State University, 1981), 100.

72. James M. Clifton, "The Rice Industry in Colonial America," *Agricultural History* 55, no. 3 (1981), 270.

73. Lefler, *New Voyage to Carolina*, 81.

74. Quoted in Carrier, *Beginnings of Agriculture*, 204; Henry C. Dethloff, *A History of the American Rice Industry, 1685–1985* (College Station: Texas A&M University Press, 1988), 10.

75. Judith A. Carney, *Black Rice: The African Origins of Rice Cultivation in the Americas* (Cambridge, MA: Harvard University Press, 2001).

76. Wood, *Black Majority*, 25, 36, 131.

77. Peter Collinson, "Of the Introduction of Rice and Tar in Our Colonies," *Gentleman's Magazine*, May 26, 1766, 278–80, quoted in A. S. Salley, "Introduction of Rice into South Carolina," *Bulletin of the Historical Commission of South Carolina* (Columbia), no. 6 (1919): 15.

78. Turner, *Africanisms*, 128.

79. McWilliams, *Revolution in Eating*, 131–65.

9. AFRICAN ANIMALS AND GRASSES
IN THE NEW WORLD TROPICS

Epigraphs: Alfred W. Crosby, *The Columbian Exchange: Biological and Cultural Consequences of 1492* (Westport, CT: Greenwood Press, 1972), 92, 95; Wesley Frank Craven, "An Introduction to the History of Bermuda," *William and Mary Quarterly Historical Magazine*, 2nd series, vol. 17, no. 3 (1937): 361.

1. Valentim Fernandes, *Description de la côte occidentale d'Afrique*, translation and notes by Th. Monod, A. Teixeira da Mota, and Raymond Mauny (Bissau, Guinea Bissau: Centro de Estudos de Guiné Portuguêsa, 1951), 28–29; Michelburne, quoted in Thomas Astley, *A New General Collection of Voyages and Travels Consisting of the Most Esteemed Relations, which Have Been hitherto Published in Any Language*, 4 vols. (London: Frank Cass, 1968), 1:307.

2. The Jobson and Lemos Coelho accounts are in David P. Gamble and P. E. H. Hair, eds., *The Discovery of River Gambra (1623) by Richard Jobson* (London: Hakluyt Society, 1999), 96 ("beeves ... graine"), 163 ("even to us ... sustenance"), 162 (on palm wine), 107 (on beer), 300 (on food animals and animal hides). Lemos Coelho also noted that the Gambia trade with the English included the hides of wild animals (294).

3. Antonio Rumeu de Armas, *España en el Africa Atlántica*, 2 vols. (Madrid: Instituto de Estudios Africanos, Consejo Superior de Investigaciones Científicas, 1956), 2:73–74, 115, 150.

4. Linda A. Newson and Susie Minchin, *From Capture to Sale: The Portuguese Slave Trade to Spanish South America in the Early Seventeenth Century* (Boston: Brill, 2007), 15–16, 84–85.

5. P. E. H. Hair, Adam Jones, Robin Law, *Barbot on Guinea: The Writings of Jean Barbot on West Africa, 1678–1712*, 2 vols. (London: Hakluyt Society, 1992), 1:45.

6. Ibid., 1:71.

7. Heywood and Thornton report the delivery of livestock—including one thousand head of cattle—to Luanda during one minor war. See Linda M. Heywood and

John K. Thornton, *Central Africans, Atlantic Creoles, and the Foundation of the Americas, 1585–1660* (New York: Cambridge University Press, 2007), 116.

8. Leif Svalesen, *The Slave Ship Fredensborg* (Bloomington: Indiana University Press, 2000), 99, 111; Robert Harms, *The Diligent: A Voyage through the Worlds of the Slave Trade* (New Haven: Yale University Press, 2002), 110; Marcus Rediker, *The Slave Ship: A Human History* (New York: Viking, 2007), 198.

9. Barbot described some of these native cattle as black and horned. Hair et al., *Barbot on Guinea*, 1:43, 66.

10. Svalesen, *Slave Ship Fredensborg*, 106, 132.

11. Guinea hens had become a novelty animal in England during the seventeenth century, introduced by ships returning from Guinea. R. A. Donkin, *Meleagrides: An Historical and Ethnogeographical Study of the Guinea Fowl* (London: Ethnographica, 1991), 84–85; Gamble and Hair, *Discovery of River Gambra*, 177, 309.

12. Astley, *New General Collection*, 2:359–60.

13. Hair et al., *Barbot on Guinea*, 1:79.

14. Father Margat, mentioned in Donkin, *Meleagrides*, 97. On the West Indies, see also David Watts, *The West Indies: Patterns of Development, Culture and Environmental Change Since 1492* (Cambridge: Cambridge University Press, 1987), 157, 198, 408.

15. Marcgraf died in Guinea on his return voyage to Europe. Jorge Marcgrave, *História natural do Brasil* (São Paulo: Imprensa Oficial do Estado, 1942), 192.

16. The information on Monticello is from the late Karen Hess, food historian, pers. comm., February 2, 2004. See also Crosby, *Columbian Exchange*, 96; Arno Vogel, Marco Antonio da Silva Mello, and José Flávio Pessoa de Barros, *A galinha d'angola: Iniciação e indentidade na cultura Afro-Brasileira* (Rio de Janeiro: Universidade Federal Fluminense, 1993).

17. The African introductions were the woolless blackbelly sheep and the smooth-haired hornless breed with shorter tails known as Persian sheep. David R. Harris, *Plants, Animals, and Man in the Outer Leeward Islands, West Indies: An Ecological Study of Antigua, Barbuda, and Anguilla,* University of California Publications in Geography 18 (Berkeley: University of California Press, 1965), 65.

18. Lemos Coelho, quoted in Gamble and Hair, *Discovery of River Gambra*, 300.

19. Sloane mentioned in Bryan Edwards, *History, Civil and Commercial, of the British Colonies in the West Indies,* 5 vols. (Chestnut Hill, MA: Adamant Media, 2001 [1793]), 1:254; quote in ibid., 1:254. See also Richard Lydekker, *The Sheep and Its Cousins* (London: G. Allen, 1912), 221.

20. Richard Ligon, *A True and Exact History of the Island of Barbadoes* (London: Frank Cass & Co., 1970 [ca. 1647]), 59. *Binny* refers to the Bight of Benin or Nigeria in our own time.

21. The African sheep breed is also known as the Barbados Blackbelly. W. R. Buttenshaw, "Barbados Woolless Sheep," *West Indian Bulletin* 6, no. 2 (1905): 187–97, esp. 187; Griffith Hughes, *The Natural History of Barbados* (New York: Arno Press, 1972), 63; Marcgrave, *História natural,* 234; Watts, *West Indies,* 164, 199; D. R. Harris, *Plants,*

Animals, and Man, 65; Food and Agriculture Organization (FAO), *Prolific Tropical Sheep,* FAO Animal Production and Health Paper, no. 17 (Rome: FAO, 1980).

22. Marcgrave, *História natural,* 230, 234.

23. Crosby, *Columbian Exchange,* 74.

24. Edmundo Wernicke, "Rutas y etapas de la introducción de los animales domésticos en las tierras americanas," *GAEA: Anales de la Sociedad Argentina de Estudios Geográficos* 6 (1938): 77–83.

25. Bahian landowner Gabriel Soares de Sousa, writing in 1584, in F. C. Hoehne, *Botánica e agricultura no Brasil no século XVI* (São Paulo: Companhia Editora Nacional, 1937), 179; Orlando Ribeiro, *Aspectos e problemas da expansão portuguésa* (Lisbon: Estudos de Ciencias Políticas e Sociais, Junta de Investigações do Ultramar, 1962), 140; T. B. Duncan, *Atlantic Islands: Madeira, the Azores, and the Cape Verdes in Seventeenth Century Commerce and Navigation* (Chicago: University of Chicago Press, 1972).

26. Ligon, *True and Exact History,* 18.

27. Alfred W. Crosby, *Ecological Imperialism: The Biological Expansion of Europe, 900–1900* (New York: Cambridge University Press, 1986), 177.

28. Ligon, *True and Exact History,* 58.

29. Ligon was in Barbados from 1647 to 1650. The map was originally drawn ca. 1640 by Captain John Swan, one of island's first English settlers (from 1628). Ligon likely added the drawings of the camels, horsemen, dolphins, and sailing ships. P. F. Campbell, "Ligon's Map," *Journal of the Barbados Museum and Historical Society* 34, no. 4 (1973): 108–12.

30. James E. McClellan III, *Colonialism and Science: Saint Domingue in the Old Regime* (Baltimore: John Hopkins University Press, 1992), 33.

31. Crosby, *Columbian Exchange,* 96; Miguel de Asúa and Roger French, *A New World of Animals: Early Modern Europeans on the Creatures of Iberian America* (Burlington, VT: Ashgate, 2005), 48–49.

32. Roger M. Blench, *Archaeology, Language, and the African Past* (Lanham, MD: Altamira Press, 2006), 251.

33. Rumeu de Armas, *España en el Africa Atlántica,* 1:56, 58, 115, 150; John Mercer, *Canary Islands: Fuerteventura* (Harrisburg, PA: Stackpole Books, 1973), 82–84, 87, and see images p. 107.

34. Figure 9.7 shows a donkey standing in the camel's shadow. One breed of Africa's indigenous donkey introduced to the Canaries and then the Azores was especially prized for its sure-footedness and its ability to carry heavy loads. In seventeenth-century Barbados these donkeys were known by the name *assinigoes.* The name likely derives from the ethnonym *Azenegues* for the Berber people who occupied the region between southern Morocco and Mauritania at the time of Iberian maritime expansion. The Azenegues traded with the Portuguese and their European successors at Arguim fort (see figure 2.3). Ligon, *True and Exact History,* 56–59; Rumeu de Armas, *España en el Africa,* 1:2, 27; Ivana Elbl, "The Horse in Fifteenth-Century Senegambia," *International Journal of African Historical Studies* 24, no. 1 (1991): 85–110.

35. James J. Parsons, "The 'Africanization' of the New World Tropical Grasslands," *Tübinger Geographische Studien* 34 (1970): 153 (first quote), 141 (second quote). In fact, it was in Brazil where many of these grasses were first collected and classified by botanists. Warren Dean, *With Broadax and Firebrand: The Destruction of the Brazilian Atlantic Forest* (Berkeley: University of California Press, 1995), 113.

36. James J. Parsons, "Spread of African Pasture Grasses to the American Tropics," *Journal of Range Management* 25 (1972): 17.

37. Richard P. Tucker, *Insatiable Appetite: The United States and the Ecological Degradation of the Tropical World* (Berkeley: University of California Press, 2000), 303; Agnes Chase, "Grasses of Brazil and Venezuela," *Agriculture in the Americas* 4 (1944): 123–26; Parsons, "Spread of African Pasture Grasses," 12–13.

38. Edwards, *History, Civil and Commercial*, 1:254.

39. Ibid., 1:253.

40. Ibid., 1:254.

41. Quoted in Peter H. Wood, *Black Majority* (New York: Knopf, 1974), 120n92.

42. Parsons, "Africanization," 145; Watts, *West Indies*, 389, 428. Besides its ubiquitous presence as pasture forage and fodder, guinea grass is used to prepare medicinal decoctions for teas. Several of the African grass introductions are used in this way in Caribbean societies. Edward S. Ayensu, *Medicinal Plants of the West Indies* (Algonac, MI: Reference Publications, 1981), 104.

43. The first quote is from Hans Sloane, *A Voyage to the Islands of Madera, Barbados, Nieves, S. Christophers and Jamaica,* 2 vols. (London: British Museum, 1707), 1:106; the second quote is from Parsons, "Spread of African Pasture Grasses," 13.

44. Edwards, *History, Civil and Commercial*, 1:253–54; Chase, "Grasses of Brazil and Venezuela," 123–26.

45. Johann Baptist von Spix and Carl Friedrich Philipp von Martius, *Travels in Brazil, in the Years 1817–1820,* 2 vols. (London: Longman, Hurst, Rees, Orme, Brown, and Green, 1824), 1:176.

46. Jean Baptiste Debret, *Viagem pitoresca e histórica ao Brasil: 1834–1839,* 2 vols. (São Paulo: Livraria Martins, 1940), 1:180–81.

47. On Cuban usage of the term *maloja,* we are indebted to Raul A. Fernandez, professor of social science, University of California, Irvine, pers. comm., April 13, 2006.

48. H. O. Neville, "The Cattle Industry of Cuba," *Cuba Review and Bulletin* (1920): 13.

49. Craven, "Introduction to the History of Bermuda"; Wernicke, "Rutas y etapas"; Gary S. Dunbar, "Colonial Carolina Cowpens," *Agricultural History* 35, no. 3 (1961): 125–31, esp. 127.

50. Bermuda grass withstands great summer heat and grows in almost any soil, especially sandy ones. It can be used for forage or cut for hay. Lewis Gray, *History of Agriculture in the Southern United States to 1860,* 2 vols. (Gloucester, MA: Peter Smith, 1958), 2:823–24; A. A. Hanson, *Grass Varieties in the United States,* Agriculture Handbook No. 170 (Washington, D.C.: USDA, 1972), 41.

51. A legume, the cowpea contributes nitrogen to soils. Riva Berleant-Schiller and Lydia M. Pulsipher, "Subsistence Cultivation in the Caribbean," *New West Indian Guide* 60, nos. 1–2 (1986): 1–40; Dunbar, "Colonial Carolina Cowpens," 126.

52. The quote is from Mark Catesby, *The Natural History of Carolina, Florida, and the Bahama Islands,* 2 vols. (London, 1771 [1743]), 1:xviii. See also Sloane, *Voyage to the Islands,* 1:104, 2:360; and Edward Long, *History of Jamaica,* 3 vols. (London: Lowndes, 1774), 2:762.

53. Quoted in Crosby, *Ecological Imperialism,* 179.

54. Quoted in J. S. Otto and Nain E. Anderson, "The Origins of Southern Cattle-Grazing: A Problem in West Indian History," *Journal of Caribbean History* 21, no. 2 (1988): 138–53, 139.

55. Gray, *History of Agriculture,* 1:19, 55; Dunbar, "Colonial Carolina Cowpens," 127; James E. McWilliams, *A Revolution in Eating: How the Quest for Food Shaped America* (New York: Columbia University Press, 2005), 133–36; Otto and Anderson, "Origins of Southern Cattle-Grazing," 138.

56. Kenneth R. Andrews, *The Spanish Caribbean: Trade and Plunder 1530–1630* (New Haven: Yale University Press, 1978), 13.

57. The Spanish fleet of 1587 carried 35,444 hides from the port of Santo Domingo to Spain; an additional 64,350 were shipped from New Spain. Harold V. Livermore, *Royal Commentaries of the Incas and General History of Peru,* 2 vols. (Austin: University of Texas Press, 1966), 1:583.

58. Clarissa Thérèse Kimber, *Martinique Revisited: The Changing Plant Geographies of a West Indian Island* (College Station: Texas A&M University Press, 1988), 124–25; McClellan, *Colonialism and Science,* 33, 69; Alexander von Humboldt, *Personal Narrative of Travels to the Equinoctial Regions of America, During the Years 1799–1804,* 3 vols. (London: Henry G. Bohn, 1853), 3:19, 90–91; James H. Sweet, *Recreating Africa: Culture, Kinship, and Religion in the African-Portuguese World, 1441–1770* (Chapel Hill: University of North Carolina Press, 2003), 38, 237n35; Dean, *With Broadax and Firebrand,* 111–14, 202–4, 272.

59. A similar point is made for enslaved Africans with prior skills and knowledge in indigo dyeing, who contributed to the crop's export development in Anglo-America. See Frederick Knight, "In an Ocean of Blue: West African Indigo Workers in the Atlantic World to 1800," in Michael A. Gomez, ed., *Diasporic Africa: A Reader* (New York: New York University, 2006), 28–44.

60. Francisco de Lemos Coelho, *Description of the Coast of Guinea, 1684,* trans. P. E. H. Hair (Liverpool: University of Liverpool, 1985), 15–16.

61. Harms, *Diligent,* 355.

62. The Spanish word for stray cattle is *cimarrón.* This word was also used for runaway slaves and is the origin of the word "maroon." John Thornton, *Africa and Africans in the Making of the Atlantic World, 1400–1680* (New York: Cambridge University Press, 1992), 135.

63. Peter H. Wood, "'It Was a Negro Taught them': A New Look at African Labor

in Early South Carolina," *Journal of Asian and African Studies* 9 (1974), 160–79, esp. 168–73; Dunbar, "Colonial Carolina Cowpens," 130; Terry G. Jordan, *Trails to Texas: Southern Roots of Western Cattle Ranching* (Lincoln: University of Nebraska Press, 1981); Andrew Sluyter, "The Role of Black Barbudans in the Establishment of Open-Range Cattle Herding in the Colonial Caribbean and South Carolina," *Journal of Historical Geography* 35 (2009): 330–49.

64. The hardy dwarf landrace of cattle found in tropical and subtropical America was known as *criollo*. It was derived from many types, all assumed to be of Iberian lineage. Potential African livestock contributions to the Moorish influence on Peninsular cattle breeding are largely unexamined. Otto and Anderson, "Origins of Southern Cattle-Grazing," 138–53; John E. Rouse, *The Criollo: Spanish Cattle in the Americas* (Norman: University of Oklahoma Press, 1977).

65. Wood, "'It Was a Negro Taught them'"; Thornton, *Africa and Africans,* 135.

66. Fulbe is the self-name for the herding people of West Africa. The Mandinka refer to them as "Fula," a form also used in English. But other terms for the Fulbe include "Fulani" or "Peul" in French.

67. Gamble and Hair, *Discovery of River Gambra,* 100.

68. Ibid., 101.

69. Barbot, quoted in Hair et al., *Barbot on Guinea,* 1:71.

70. Francis Moore, *Travels into the Inland Parts of Africa* (London: Edward Cave, 1738), 24.

71. Gamble and Hair, *Discovery of River Gambra,* 66–67.

72. Moore, *Travels,* 24.

73. Lemos Coelho, quoted in Gamble and Hair, *Discovery of River Gambra,* 294.

74. George Richardson Porter, *The Nature and Properties of the Sugar Cane: With Practical Directions for the Improvement of Its Culture and the Manufacture of Its Products* (London: Smith, Elder, and Co., 1830), 40.

75. Buttenshaw, "Barbados Woolless Sheep," esp. 188, 195; Harris, *Plants, Animals, and Man,* 65.

76. Berleant-Schiller and Pulsipher, "Subsistence Cultivation," 15; Riva Berleant-Schiller, "The Social and Economic Role of Cattle in Barbuda," *Geographical Review* 67, no. 3 (1977): 299–309.

77. On the cowpen system for fertilizing fallow sugarcane fields in the British West Indies during the plantation era, see Barry Higman, *Slave Populations of the British Caribbean, 1807–1834* (Kingston, Jamaica: University of the West Indies, 1995), 164. On the cultural interchange that resulted in the penning technique, see Otto and Anderson, "Origins of Southern Cattle-Grazing."

78. Riva Berleant-Schiller, "Grazing and Gardens in Barbuda," in Riva Berleant-Schiller and Eugenia Shanklin, eds., *The Keeping of Animals: Adaptation and Social Relations in Livestock Producing Communities* (Totowa, NJ: Allanheld and Osmun, 1983), 73–91; Sluyter, "Role of Black Barbudans."

79. Otto and Anderson, "Origins of Southern Cattle-Grazing," 146.

80. Wood, *Black Majority*, 131; Otto and Anderson, "Origins of Southern Cattle-Grazing," 147–48.

81. K. G. Davies, *The Royal African Company* (New York: Atheneum, 1970 [1957]), 44–45.

82. Lemos Coelho, *Description of the Coast of Guinea*, 14–16; Lemos Coelho, quoted in Gamble and Hair, *Discovery of River Gambra*, 294.

83. Craven, "Introduction to the History of Bermuda," first quote 361; second quote 354; third quote 361.

10. MEMORY DISHES OF THE AFRICAN DIASPORA

Epigraph: Sidney W. Mintz, professor emeritus of anthropology, John Hopkins University, pers. comm., May 27, 2008.

1. James E. McWilliams, *A Revolution in Eating: How the Quest for Food Shaped America* (New York: Columbia University Press, 2005), 11–13.

2. F. R. Irvine, "The Edible Cultivated and Semi-cultivated Leaves of West Africa," *Qualitas Plantarum et Materiae Vegetabiles*, no. 2 (1956): 35–42.

3. This is the plant identified by "F" in his drawing. Pieter de Marees, *Description and Historical Account of the Gold Kingdom of Guinea (1602)*, trans. and ed. Albert van Dantzig and Adam Jones (Oxford: Oxford University Press, 1987), 158. A dish made with *nkasa* leaves was described in one seventeenth-century account of Luanda. Linda M. Heywood and John K. Thornton, *Central Africans, Atlantic Creoles, and the Foundation of the Americas, 1585–1660* (New York: Cambridge University Press, 2007), 216. Moore wrote of two leafy vegetables, one he called "purselain" that grew near the Gambia River, noting "it is very good, resembling the *English.*" Francis Moore, *Travels into the Inland Parts of Africa* (London: Edward Cave, 1738), 62.

4. James A. Chweya and Pablo B. Eyzaguirre, *Biodiversity of Traditional Leafy Vegetables* (Rome: IPGRI, 1999), 1–2.

5. The early nineteenth-century slave trader Theophilus Conneau remarked on the rich stews he was served in Guinea. Theophilus Conneau, *A Slaver's Log Book or 20 Years' Residence in Africa* (Englewood Cliffs, NJ: Prentice-Hall, 1976), 83–84, 95. On the African stews, see Jessica B. Harris, *The Africa Cookbook* (New York: Simon and Schuster, 1998).

6. On the preference for the bitter trait in African cooking, see National Research Council (NRC), *Lost Crops of Africa*, vol. 2, *Vegetables* (Washington, D.C.: National Academy Press, 2006), 140.

7. Ibid., 140; Jessica B. Harris, *Iron Pots and Wooden Spoons* (New York: Atheneum Press, 1989), 28; Dorothea Bedigian, "Slimy Leaves and Oily Seeds: Distribution and Use of Wild Relatives of Sesame in Africa," *Economic Botany* 58 (supplement) (2004): 164–94; H. M. Burkill, *The Useful Plants of West Tropical Africa*, 6 vols. (Kew, England: Royal Botanic Gardens, 1985–2004), 4:37.

8. Chweya and Eyzaguirre, *Biodiversity;* Franklin W. Martin, Ruth M. Ruberté, Laura S. Meitzner, *Edible Leaves of the Tropics* (North Fort Myers, FL: Echo Press, 1998).

9. Manioc leaves are only occasionally used in Amerindian societies, where they have been of far less dietary consequence. In Cameroon, an alcoholic drink is made by fermenting manioc pulp. W. O. Jones, "Manioc: An Example of Innovation in African Economies," *Economic Development and Cultural Change* 5, no. 2 (1957): 97–117, esp. 113; Ifeyironwa Francisca Smith, *Food of West Africa: Their Origin and Use* (Ottawa: Kwik Kopy, 1998), 236; Burkill, *Useful Plants of West Tropical Africa*, 1:598.

10. W. O. Jones, "Manioc," 102. The young leaves of manioc are also eaten in some parts of Latin America. Eugene Anderson, professor emeritus of anthropology, University of California, Riverside, pers. comm., September 8, 2008.

11. Both greens were present in sub-Saharan Africa prior to the transatlantic slave trade. Collard greens (kale, or *couve* in Portuguese, *Brassica oleracea*) are of Mediterranean origin; they diffused from North Africa southward in the millennia before recorded history. Undoubtedly familiar to Iberians and southern Europeans, in the Americas collards have nevertheless been associated with diasporic foodways. Mustard greens *(Brassica juncea)* are possibly of African origin but may have arrived there from Asia in prehistory. Burkill, *Useful Plants of West Tropical Africa,* 1:560–62; Kenneth Kiple and Kriemheld Ornelas, eds., *The Cambridge World History of Food,* 2 vols. (Cambridge: Cambridge University Press, 2000) 2:1761, 1819–20. See also Elizabeth Schneider, *Vegetables from Amaranth to Zucchini* (New York: William Morrow, 2001), 211–12, 396; Luis da Camara Cascudo, *História da alimentação no Brasil,* 2 vols. (São Paulo: Editora Itatiaia, 1983), 1:245; and NRC, *Lost Crops,* 2:143.

12. In Brazilian candomblé, sorrel is also prepared as an offering to the African deity Shango (Xangô). José Flávio Pessoa de Barros and Eduardo Napoleão, *Ewé òrìsà: Uso litúrgico e terapêutico dos vegetais nas casas de candomblé jêje-nago* (Rio de Janeiro: Betrand Brasil, 1998), 174. See also Burkill, *Useful Plants of West Tropical Africa,* 4:36; Walter B. Mors, Carlos Toledo Rizzini, and Nuno Álvares Pereira, *Medicinal Plants of Brazil* (Algonac, MI: Reference Publications, 2000), 221; and Edward S. Ayensu, *Medicinal Plants of the West Indies* (Algonac, MI: Reference Publications, 1981), 120.

13. *Vernonia amygdalina* and to some extent *V. colorata* are favored in African foodways. Chweya and Eyzaguirre, *Biodiversity,* 4; Burkill, *Useful Plants of West Tropical Africa,* 1:501–3; Kay Williamson, "Food Plant Names in the Niger Delta." *International Journal of American Linguistics* 36, no. 2 (1970): 156–67, esp. 161; Martin et al., *Edible Leaves of the Tropics,* 37. *Vernonia amygdalina* was likely introduced to Jamaica during the slave trade; it is a longstanding herbal tea for colic. Ayensu, *Medicinal Plants of the West Indies,* 84. The African species is also featured in Brazilian candomblé. Barros and Napoleão, *Ewé òrìsà,* 83.

14. G. J. M. Grubben and O. A. Denton, eds., *Plant Resources of Tropical Africa,* vol. 2, *Vegetables* (Leiden: Backhuys, 2004); NRC, *Lost Crops,* 2:35–41; Martin et al., *Edible Leaves of the Tropics,* 10.

15. Moore, *Travels into the Inland,* 108.

16. NRC, *Lost Crops,* 2:37n5. In Brazil the amaranth spinach is known as *bredo.*

17. Arnold R. Highfield and Vladimir Barac, *C. G. A. Oldendorp's History of the Mission of the Evangelical Brethren on the Caribbean Islands of St. Thomas, St. Croix, and St. John* (Ann Arbor, MI: Karoma Publishers, 1987), 660n31.

18. John T. Schneider, *Dictionary of African Borrowings in Brazilian Portuguese* (Hamburg: Helmut Buske Verlag, 1991), 109–10; F. G. Cassidy and R. B. Le Page, *Dictionary of Jamaican English* (Kingston, Jamaica: University of the West Indies Press, 2002), 89. In Haiti the term *callaloo* refers to okra; in other parts of the Caribbean *callalou* is made with taro leaves. In Brazil *caruru-azedo* refers to African *Hibiscus sabdariffa*.

19. Harris, *Iron Pots and Wooden Spoons;* Cascudo, *História da Alimentação.*

20. Quoted in Ray Kea, *Settlements, Trade, and Polities in the Seventeenth-Century Gold Coast* (Baltimore: Johns Hopkins University Press, 1982), 301. The mention of potatoes could be a reference to the native African dazo potato *(Plectranthus esculentus)* or the New World sweet potato.

21. Jamaican slaves referred to the dish as *foo-foo.* It was made from yams, plantains, or cassava. Barry Higman, *Slave Populations of the British Caribbean, 1807–1834* (Kingston, Jamaica: University of the West Indies, 1995), 216.

22. Today the Brazilian dish is typically prepared from maize and known as *mingau.* Jean Baptiste Debret, *Viagem pitoresca e histórica ao Brasil: 1834–1839,* 2 vols. (São Paulo: Livraria Martins, 1940 [1834–39]), 1:228–29; Cascudo, *História da alimentação,* 1:222.

23. Marees, *Description and Historical Account,* 62.

24. Heywood and Thornton, *Central Africans, Atlantic Creoles,* 216.

25. Han Jordaan, "The Curaçao Slave Market: From *Asiento* Trade to Free Trade, 1700–1730," in Postma and Enthoven, *Riches from Atlantic Commerce,* 219–57, esp. 237.

26. Highfield and Barac, *Oldendorp's History,* 107.

27. The many "rice and pea" dishes of the Caribbean are often prepared with African pigeon peas, such as *arroz con guandules* of Puerto Rico. These rice and bean dishes incidentally provide most of the dietary protein required for human health. The Louisiana jambalaya resulted from cooking raw rice in the broth of a stew composed of ingredients from many cultural heritages. Anthropologist Eugene Anderson argues that jambalaya is a variation of the Senegalese dish of Jollof (Wolof) rice, also known as *ceebu jen* in Wolof. He adds that the word *jamba* in Wolof means "mixed," while *laaya* refers to a type of stew. Eugene Anderson, pers. comm., September 5, 2008.

28. Burkill, *Useful Plants of West Tropical Africa,* 4:36; Mors et al., *Medicinal Plants of Brazil,* 221.

29. Also included in Rutledge's cookbook are recipes for benne (sesame) soup and groundnut soup, the latter a signature West African stew made from peanuts. Karen Hess, *The Carolina Rice Kitchen* (Columbia: University of South Carolina Press, 1992), 102.

30. Woodville K. Marshall, "Provision Ground and Plantation Labor in Four Windward Islands: Competition for Resources during Slavery," in Berlin and Morgan, *Cul-*

tivation and Culture, 203–20, esp. 212; Higman, *Slave Populations,* 241; Michael Mullin, *Africa in America* (Urbana: University of Illinois Press, 1994), 132, 304–7; Mary Karasch, "Slave Women on the Brazilian Frontier in the Nineteenth Century," in Gaspar and Hine, *More Than Chattel,* 79–96, esp. 87, 95n23; David Barry Gaspar, "From 'The Sense of Their Slavery': Slave Women and Resistance in Antigua, 1632–1763," in Gaspar and Hine, *More Than Chattel,* 218.

31. On *karité (gharti),* see Tadeusz Lewicki, *West African Food in the Middle Ages* (Cambridge: Cambridge University Press, 1974), 36; N. Levtzion and J. F. P. Hopkins, *Corpus of Early Arabic Sources for West African History* (Princeton, NJ: Marcus Wiener Publishers, 2000), 287.

32. J. T. Schneider, *Dictionary of African Borrowings,* 129.

33. There is no evidence from pre-Columbian history that Native Americans cooked their food by frying with vegetable oil, even though tropical America abounds in many types of oil-bearing nuts, seeds, and palm trees. It is thought that the lack of advanced technologies for extraction and use inhibited oil production in dependable quantities. Significantly, in the Americas there were no iron implements—with their high heat-conducting properties for deep-frying—until the arrival of Europeans and Africans skilled in blacksmithing. In West Africa and the Mediterranean, each with an ancient iron-smithing tradition, food is traditionally cooked in vegetable oil; however, deep-frying is far more common in African cookery. A vegetable-oil-based cooking tradition was certainly evident in the seventeenth-century quilombo Palmares. The report of Captain Johan Blaer, leader of the Dutch militia sent against it in 1645, mentioned that his men found iron implements, blacks with iron-forging skills, and stores of palm oil used for cooking. As most of the quilombolas in Dutch Brazil at this time were Africa-born, Blaer's observations suggest a perpetuation of African cooking practices among the runaways. Adelmir Fiabani, *Mato, Palhoça, e Pilão: O quilombo, da escravidão às comunidades remanescentees [1532–2004]* (São Paulo: Editora Expressão Popular, 2005), 325, 341; Sophie D. Coe, *America's First Cuisines* (Austin: University of Texas Press, 1994), 36; Linda A. Newson and Susie Minchin, *From Capture to Sale: The Portuguese Slave Trade to Spanish South America in the Early Seventeenth Century* (Boston: Brill, 2007), 173, 304; Eugene Anderson, pers. comm., September 8, 2008.

34. Patricia L. Howard, "Women and the Plant World: An Exploration," in Patricia L. Howard, ed., *Women and Plants: Gender Relations in Biodiversity Management and Conservation* (London: Zed, 2003), 11.

35. The chronic hunger and privation that typified the early period of plantation slavery persisted through much of the eighteenth century on Caribbean islands, where sugar monoculture left little land for subsistence. Themes of memory, slavery, hunger, and food are features of some contemporary Caribbean writing. For example, diaspora women are linked to their African counterparts through the cultivation of yams in Kamau Brathwaite's poem "The Dust." Edward [Kamau] Brathwaite, "The Dust," *Rights of Passage* (London: Oxford University Press, 1967), 65. George Lamming's novel *In the Castle of My Skin* signifies impending cultural dislocation through a farewell dish

of okra and flying fish, which a mother prepares for her son on the eve of his migration. George Lamming, *In the Castle of My Skin* (London: Longman, 1970). Two journals that focus on the Caribbean also employ food in their titles. *Callaloo* is the name of the leading journal of arts, letters, and cultures of the African diaspora. *Cajanus,* named after the genus of the pigeon pea, is the Caribbean journal of food and nutrition. On Brathwaite and Lamming, also see the essay by Louis James, "Eating the Dead," in Hena Maes-Jelinek, Gordon Collier, Geoffrey V. Davis, eds., *A Talent(ed) Digger: Creations, Cameos, and Essays in Honour of Anna Rutherford* (Atlanta: Rodopi, 1996), 309–12.

36. Other African plants featured in candomblé ceremonies include oil palm, sesame, kola, melegueta and guinea pepper, watermelon, jute mallow, amaranth, hibiscus, and castor bean. Robert A. Voeks, *Sacred Leaves of Candomblé* (Austin: University of Texas Press, 1997), 78–79; Barros and Napoleão, *Ewé òrìsà.*

SELECTED BIBLIOGRAPHY

Abaka, Edmund. "Kola Nut." In Kiple and Ornelas, *Cambridge World History of Food,* 1:684–92.

Agorsah, E. Kofi. "Scars of Brutality: Archaeology of the Maroons in the Caribbean." In Ogundiran and Falola, *Archaeology of Atlantic Africa,* 332–54.

Alencastro, Luiz Felipe de. *O trato dos viventes: Formação do Brasil no Atlântico Sul; Séculos XVI e XVII.* São Paulo: Companhia das Letras, 2000.

Alexander, J., and D. G. Coursey. "The Origins of Yam Cultivation." In P. J. Ucko and G. W. Dimbleby, eds., *The Domestication and Exploitation of Plants and Animals,* 405–25. London: Gerald Duckworth, 1969.

Almada, André Álvares de. *Brief Treatise on the Rivers of Guinea (c. 1594).* Liverpool: Department of History, University of Liverpool, 1984.

Alpern, Stanley B. "The European Introduction of Crops into West Africa in Precolonial Times." *History in Africa* 19 (1992): 13–43.

Andrews, Kenneth R. *The Spanish Caribbean: Trade and Plunder 1530–1630.* New Haven: Yale University Press, 1978.

Anon. "Het guineesche graan." *Curaçaosche Courant,* January 20, 1838.

Anon., [ed.]. "Origin of the Banana." *Journal of Heredity* 5, no. 6 (1914): 273–80.

Araújo dos Anjos, Rafael Sanzio. *Territórios das comunidades quilombolas do Brasil.* Brasilia: Mapas Editora & Consultoria, 2005.

Arbell, M. *The Jewish Nation of the Caribbean.* Jerusalem: Gefen Publishing House, 2002.

Assadourian, Carlos Sempat. *El tráfico de esclavos en Córdoba de Angola a Potosí, siglos XVI–XVII.* Córdoba: Universidade Nacional de Córdoba, 1966.

Astley, Thomas. *A New General Collection of Voyages and Travels: Consisting of the Most Esteemed Relations, which Have Been hitherto Published in Any Language.* 4 vols. London: Thomas Astley, 1745–47; London: Cass, 1968.

Asúa, Miguel de, and Roger French. *A New World of Animals: Early Modern Europeans on the Creatures of Iberian America.* Burlington, VT: Ashgate, 2005.

Ayensu, Edward S. *Medicinal Plants of the West Indies.* Algonac, MI: Reference Publications, 1981.

———. *Medicinal Plants of West Africa.* Algonac, MI: Reference Publications, 1978.

Barbot, Jean. "A Description of the Coasts of North and South Guinea; and of Ethiopia Inferior, Vulgarly Angola; Being a New and Accurate Account of the Western Maritime Countries of Africa." In Awnsham Churchill, *A Collection of Voyages and Travels, Some Now First Printed from Original Manuscripts, Others now First Published in English,* 8 vols., 5:1–522. London: Printed from Messieurs Churchill, for T. Osborne, 1752.

Barham, Henry. *Hortus Americanus.* Kingston, Jamaica: Alexander Aikman, 1794.

Barros, José Flávio Pessoa de, and Eduardo Napoleão. *Ewé òrìsà: Uso litúrgico e terapêutico dos vegetais nas casas de candomblé jêje-nago.* Rio de Janeiro: Betrand Brasil, 1998.

Barry, Boubacar. *Senegambia and the Slave Trade.* Cambridge: Cambridge University Press, 1998.

Bedigian, Dorthea. "Sesame in Africa: Origin and Dispersals." In Neumann et al., *Food, Fuel and Fields,* 17–36.

———. "Slimy Leaves and Oily Seeds: Distribution and Use of Wild Relatives of Sesame in Africa." *Economic Botany* 58 (supplement) (2004): 164–94.

Berleant-Schiller, Riva. "Grazing and Gardens in Barbuda." In Riva Berleant-Schiller and Eugenia Shanklin, eds., *The Keeping of Animals: Adaptation and Social Relations in Livestock Producing Communities,* 73–91. Totowa, NJ: Allanheld and Osmun, 1983.

———. "Hidden Places and Creole Forms: Naming the Barbudan Landscape." *Professional Geographer* 43, no. 1 (1991): 92–101.

———. "The Social and Economic Role of Cattle in Barbuda." *Geographical Review* 67, no. 3 (1977): 299–309.

Berleant-Schiller, Riva, and Lydia M. Pulsipher. "Subsistence Cultivation in the Caribbean." *New West Indian Guide* 60, nos. 1–2 (1986): 1–40.

Berlin, Ira, and Philip D. Morgan, eds. *Cultivation and Culture: Labor and the Shaping of Slave Life in the Americas.* Charlottesville: University Press of Virginia, 1993.

———. Introduction to Berlin and Morgan, *Cultivation and Culture,* 1–45.

Besson, Jean. "Family Land and Caribbean Society: Toward an Ethnography of Afro-Caribbean Peasantries." In E. Thomas-Hope, ed., *Perspectives on Caribbean Regional Identity,* 57–83. Liverpool: Centre for Latin American Studies, University of Liverpool, 1984.

Betts, Edwin Morris, ed. *Thomas Jefferson's Garden Book, 1766–1824.* Philadelphia: American Philosophical Society, 1944.

Biblioteca Nacional. *Anais da Biblioteca Nacional,* vol. 108 of 1988. Rio de Janeiro: Fundação Biblioteca Nacional, 1992.

Bilby, Kenneth M. *True-Born Maroons.* Gainesville: University of Florida Press, 2005.

Blench, Roger M. *Archaeology, Language, and the African Past.* Lanham, MD: Altamira Press, 2006.

———. "The Movement of Cultivated Plants between Africa and India in Prehistory." In Neumann et al., *Food, Fuel and Fields,* 273–92.

Blench, Roger M., and Kevin C. MacDonald. *The Origins and Development of African Livestock: Archaeology, Genetics, Linguistics and Ethnography.* London: University College London, 2000.

Bloom, H. *The Economic Activities of the Jews of Amsterdam in the Seventeenth and Eighteenth Centuries.* Williamsport, PA: Bayard Press, 1937.

Bosman, Willem. *A New and Accurate Description of the Coast of Guinea, Divided into the Gold, the Slave, and the Ivory Coasts.* London: Printed for James Knapton and Dan. Midwinter, 1705.

Bovill, E. W. *The Golden Trade of the Moors.* Oxford: Oxford University Press, 1968.

Bradley, D. G., D. E. MacHugh, P. Cunningham, and R. T. Loftus. "Mitochondrial Diversity and the Origins of African and European Cattle." *Proceedings of the National Academy of Sciences (PNAS)* 93, no. 10 (1996): 5131–35.

Brathwaite, Edward [Kamau]. *The Development of Creole Society in Jamaica, 1770–1820.* Oxford: Clarendon, 1971.

———. "The Dust." In *Rights of Passage.* London: Oxford University Press, 1967.

Bryant, Charles. *Flora Diaetetica: Or, History of Esculent Plants, Both Domestic and Foreign; In which They Are Accurately Described, and Reduced to Their Linnaean Generic and Specific Names, with Their English Names Annexed.* London: B. White, 1783.

Brydon, L. "Rice, Yams and Chiefs in Avatime: Speculations on the Development of a Social Order." *Africa* 51, no. 2 (1981): 659–77.

Bubberman, F. C., et al. *Links with the Past: The History of the Cartography of Suriname 1500–1971.* Amsterdam: Theatrum Orbis Terrarum B.V., 1973.

Burkill, H. I. "The Contact of the Portuguese with African Food Plants Which Gave Words Such As 'Yam' to European Languages." *Proceedings of the Linnean Society of London,* part 2 (1938): 84–95.

Burkill, H. M. *The Useful Plants of West Tropical Africa,* 6 vols. Kew, England: Royal Botanic Gardens, 1985–2004.

Burns, E. Bradford. *A History of Brazil.* New York: Columbia University Press, 1970.

Buttenshaw, W. R. "Barbados Woolless Sheep." *West Indian Bulletin* 6, no. 2 (1905): 187–97.

Buvelot, Quentin, ed. *Albert Eckhout: A Dutch Artist in Brazil.* The Hague: NIB Capital, 2004.

Cabrera, Lydia. *El Monte.* Miami: Ediciones Universal, 1995 [1954].

Campbell, Joseph. *Historical Atlas of World Mythology.* 3 vols. New York: Harper & Row, 1988.

Campbell, P. F. "Ligon's Map." *Journal of the Barbados Museum and Historical Society* 34, no. 4 (1973): 108–12.

Cardoso, Ciro Flamarion S. "The Peasant Breach in the Slave System: New Developments in Brazil." *Luso-Brazilian Review* 25, no. 1 (1988): 49–57.

Carneiro, Edison. *O Quilombo dos Palmares*. São Paulo: Companhia Editora Nacional, 1958.

Carney, Judith A. "African Traditional Plant Knowledge in the Circum-Caribbean Region." *Journal of Ethnobiology* 23, no. 2 (2003): 167–85.

———. *Black Rice: The African Origins of Rice Cultivation in the Americas*. Cambridge, MA: Harvard University Press, 2001.

———. "'With Grains in Her Hair': Rice History and Memory in Colonial Brazil," *Slavery and Abolition*, 25, no. 1 (2004): 1–27.

Carney, Judith, and Robert Voeks. "Landscape Legacies of the African Diaspora in Brazil." *Progress in Human Geography* 27, no. 2 (2003): 139–52.

Carreira, António. *As companhias pombalinas de Grão-Pará e Maranhão e Pernambuco e Paraíba*. Lisboa: Editorial Presença, 1983.

Carrier, Lyman. *The Beginnings of Agriculture in America*. New York: McGraw-Hill, 1923.

Cascudo, Luis da Camara. *História da alimentação no Brasil*. 2 vols. São Paulo: Editora Itatiaia, 1983.

Cassidy, F. G., and R. B. Le Page. *Dictionary of Jamaican English*. Kingston, Jamaica: University of the West Indies Press, 2002.

Catesby, Mark. *The Natural History of Carolina, Florida, and the Bahama Islands*. 2 vols. London, 1771 [1743].

Cavazzi de Montecúccolo, João António. *Descrição histórica dos três reinos do Congo: Matamba e Angola*. Lisboa: Junta de Investigações do Ultramar, 1965 [ca. 1660].

Chambers, Douglas B. *Murder at Montpelier: Igbo Africans in Virginia*. Jackson: University Press of Mississippi, 2005.

Chase, Agnes. "Grasses of Brazil and Venezuela." *Agriculture in the Americas* 4 (1944): 123–26.

Churchill, Awnsham. *A Collection of Voyages and Travels, Some Now First Printed from Original Manuscripts, Others now First Published in English*. 8 vols. London: Printed from Messieurs Churchill, for T. Osborne, 1752.

Chweya, James A., and Pablo B. Eyzaguirre. *Biodiversity of Traditional Leafy Vegetables*. Rome: IPGRI, 1999.

CIRAD. *Mémento de l'agronome*. Montpellier, France: CIRAD/Ministère des Affaires Étrangères, 2002.

Clifton, James M. "The Rice Industry in Colonial America." *Agricultural History* 55, no. 3 (1981): 266–83.

Clutton-Brock, Julie. *A Natural History of Domesticated Mammals*. Cambridge: Cambridge University Press, 1999.

Coe, Sophie D. *America's First Cuisines*. Austin: University of Texas Press, 1994.

Collinson, Peter. "Of the Introduction of Rice and Tar in Our Colonies." *Gentleman's Magazine,* May 26, 1766, 278–80.

Conneau, Theophilus. *A Slaver's Log Book or 20 Years' Residence in Africa.* Englewood Cliffs, NJ: Prentice-Hall, 1976.

Conrad, Robert Edgar. *The Children of God's Fire: A Documentary History of Black Slavery in Brazil.* University Park: Pennsylvania State University Press, 1984.

Coughtry, Jay. *The Notorious Triangle: Rhode Island and the African Slave Trade, 1700–1807.* Philadelphia: Temple University Press, 1981.

Counter, S. A., and D. L. Evans. *I Sought My Brother: An Afro-American Reunion.* Cambridge, MA: MIT Press, 1981.

Coursey, D. G., and C. K. Coursey. "The New Yams Festivals of West Africa." *Anthropos* 66, nos. 3–4 (1971): 444–84.

Covey, Cyclone, trans. *Cabeza de Vaca's Adventures in the Unknown Interior of America.* Albuquerque: University of New Mexico Press, 1983.

Craven, Wesley Frank. "An Introduction to the History of Bermuda," *William and Mary Quarterly Historical Magazine,* 2nd series, vol. 17, no. 3 (1937): 317–62.

Crone, Gerald Roe. *The Voyages of Cadamosto and Other Documents on Western Africa in the Second Half of the Fifteenth Century.* London: Hakluyt Society, 1937.

Crosby, Alfred W. *The Columbian Exchange: Biological and Cultural Consequences of 1492.* Westport, CT: Greenwood Press, 1972.

———. *Ecological Imperialism: The Biological Expansion of Europe, 900–1900.* New York: Cambridge University Press, 1986.

Cunha, Lygia da Fonseca Fernandes da. *Riscos iluminados de figurinhos de brancos e negros dos uzos de Rio de Janeiro e Serro do Frio, aquarelas Carlos Julião.* Rio de Janeiro: Biblioteca Nacional, 1960.

Curtin, Philip D. *The Atlantic Slave Trade.* Madison: University of Wisconsin Press, 1969.

———. *Economic Change in Pre-colonial Africa.* Madison: University of Wisconsin Press, 1975.

Curto, José C., and Paul E. Lovejoy, *Enslaving Connections: Changing Cultures of Africa and Brazil during the Era of Slavery.* Amherst, NY: Humanity Books, 2004.

Dabbs, Edith M. *Face of an Island.* Columbia, SC: R. L. Bryant, 1971.

D'Andrea, A. C., and J. Casey. "Pearl Millet and Kintampo Subsistence." *African Archaeobotanical Review* 19, no. 3 (2002): 147–73.

Davies, K. G. *The Royal African Company.* New York: Atheneum, 1970 [1957].

Davis, Nathalie Zemon. *Women on the Margins: Three Seventeenth-Century Lives.* Cambridge, MA: Harvard University Press, 1995.

Dean, Warren. *With Broadax and Firebrand: The Destruction of the Brazilian Atlantic Forest.* Berkeley: University of California Press, 1995.

de Beer, Gavin. *Hannibal: Challenging Rome's Supremacy.* New York: Viking, 1969.

Debret, Jean Baptiste. *Viagem pitoresca e histórica ao Brasil: 1834–1839.* 2 vols. São Paulo: Livraria Martins, 1940 [1834–39].

Debrien, Gabriel. "Marronage in the French Caribbean." In R. Price, *Maroon Societies*, 107–34.

Decker-Walters, Deena S., Mary Wilkins-Ellert, Sang-Min Chung, and Jack E. Staub. "Discovery and Genetic Assessment of Wild Bottle Gourd [Lagenaria Siceraria (Mol.) Standley; Cucurbitaceae] from Zimbabwe." *Economic Botany* 58, no. 4 (2004): 501–8.

de Groot, Silvia W. "Maroons of Surinam: Dependence and Independence." *Annals of the New York Academy of Sciences* 292 (1977): 455–63.

de Langhe, Edmond. *Banana and Plantain: The Earliest Fruit Crops?* Annual Report. Montpellier, France: INIBAP, 1995.

de Langhe, Edmond, R. Swennen, and D. Vuylsteke. "Plantain in the Early Bantu World." *Azania* 29–30 (1994–95): 147–60.

Déme, Alioune, and Ndèye Sokhna Guèye. "Enslavement in the Middle Senegal Valley: Historical and Archaeological Perspectives." In Ogundiran and Falola, *Archaeology of Atlantic Africa*, 122–39.

Dethloff, Henry C. *A History of the American Rice Industry, 1685–1985*. College Station: Texas A&M University Press, 1988.

de Wet, Jan M. J. "Sorghum." In Kiple and Ornelas, *Cambridge World History of Food*, 1:152–58.

Dicum, Gregory, and Nina Luttinger. *The Coffee Book: Anatomy of an Industry from Crop to the Last Drop*. New York: The New Press, 1999.

Donkin, R. A. *Meleagrides: An Historical and Ethnogeographical Study of The Guinea Fowl*. London: Ethnographica, 1991.

Donnan, Elizabeth. *Documents Illustrative of the History of the Slave Trade to America*. 4 vols. Washington, D.C.: Carnegie Institution, 1930–35.

Dow, George Francis. *Slave Ships and Slaving*. Salem, MA: Marine Research Society, 1927.

Drake, St. Clair. *Black Folk Here and There*. 2 vols. Los Angeles: UCLA/CAAS, 1990.

Dunbar, Gary S. "Colonial Carolina Cowpens." *Agricultural History* 35, no. 3 (1961): 125–31.

Duncan, T. B. *Atlantic Islands: Madeira, the Azores, and the Cape Verdes in Seventeenth Century Commerce and Navigation*. Chicago: University of Chicago Press, 1972.

Edwards, Bryan. *History, Civil and Commercial, of the British Colonies in the West Indies*. 5 vols. Chestnut Hill, MA: Adamant Media, 2001 [1793].

Ehret, Christopher. *An African Classical Age: Eastern and Southern Africa in World History, 1000 B.C. to A.D. 400*. Charlottesville: University Press of Virginia, 1998.

———. "Historical/Linguistic Evidence for Early African Food Production." In J. Desmond Clark and Steven A. Brandt, eds., *From Hunters to Farmer: The Causes and Consequences of Food Production in Africa*, 25–35. Berkeley: University of California Press, 1984.

———. "Linguistic Stratigraphies and Holocene History in Northeastern Africa." In Marek Chlodnicki and Karla Kroeper, eds., *Archaeology of Early Northeastern Africa*,

Studies in African Archaeology 9, 1019–55. Posnán, Poland: Posnán Archaeological Museum, 2006.

———. "Sudanic Civilization." In Michael Adas, ed., *Agricultural and Pastoral Societies in Ancient and Classical History,* 224–74. Philadelphia: Temple University Press, 2001.

Elbl, Ivana. "The Horse in Fifteenth-Century Senegambia." *International Journal of African Historical Studies* 24, no. 1 (1991): 85–110.

———. "'Slaves Are a Very Risky Business . . .': Supply and Demand in the Early Atlantic Slave Trade." In Curto and Lovejoy, *Enslaving Connections,* 29–55.

Eltis, David, ed. *Coerced and Free Migration: Global Perspectives.* Stanford: Stanford University Press, 2002.

———. *The Rise of African Slavery in the Americas.* Cambridge: Cambridge University Press, 2000.

Eltis, David, S. Behrendt, D. Richardson, and Herbert S. Klein, eds. *The Trans-Atlantic Slave Trade: A Database on CD-ROM.* Cambridge: Cambridge University Press, 1999.

Epstein, H. *The Origin of the Domestic Animals of Africa,* 2 vols. New York: Africana Publishing, 1971.

Erickson, D. L., Bruce D. Smith, A. C. Clarke, D. H. Sandweiss, and N. Tuross. "An Asian Origin for a 10,000-Year-Old Domesticated Plant in the Americas." *Proceedings of the National Academy of Sciences (PNAS)* 102, no. 51 (2005): 18315–20.

Esquivel, Miguel, and Karl Hammer, "The Cuban Homegarden 'Conuco': A Perspective Environment for Evolution and In Situ Conservation of Plant Genetic Resources." *Genetic Resources and Crop Evolution* 39, no. 1 (1992): 9–22.

Eyongetah, Tambi, and Robert Brain, *A History of the Cameroon.* London: Longmans, 1974.

Fage, J. D. "A New Check List of the Forts and Castles of Ghana." *Transactions of the Historical Society of Ghana* 4, no. 1 (1959): 57–67.

Falconbridge, Alexander. *An Account of the Slave Trade on the Coast of Africa.* London: J. Phillips, 1788.

Fenn, Elizabeth A. *Pox Americana.* New York: Hill and Wang, 2001.

Fernandes, Valentim. *Description de la côte occidentale d'Afrique.* Translation and notes by Th. Monod, A. Teixeira da Mota, and Raymond Mauny. Bissau, Guinea Bissau: Centro de Estudos de Guiné Portuguêsa, 1951.

Fernández-Armesto, Felipe. *Before Columbus: Exploration and Colonization from the Mediterranean to the Atlantic, 1229–1492.* Philadelphia: University of Pennsylvania Press, 1987.

Ferreira, Roquinaldo, and Flávio dos Santos Gomes. "African Diaspora Studies in/and Brazil." Paper presented at the African Diaspora Studies and the Disciplines Conference, University of Wisconsin-Madison, March 23–26, 2006.

Fiabani, Adelmir. *Mato, Palhoça, e Pilão: O quilombo, da escravidão às comunidades remanescentes [1532–2004].* São Paulo: Editora Expressão Popular, 2005.

Flory, Thomas. "Fugitive Slaves and Free Society: The Case of Brazil." *Journal of Negro History* 64, no. 2 (1979): 116–30.

Food and Agriculture Organization (FAO). *Prolific Tropical Sheep.* FAO Animal Production and Health Paper, no. 17. Rome: FAO, 1980.

Fuller, Dorian Q. "African Crops in Prehistoric South Asia: A Critical Review." In Neumann et al., *Food, Fuel and Fields,* 239–71.

Funari, Pedro P. "The Archaeological Study of the African Diaspora in Brazil." In Ogundiran and Falola, *Archaeology of Atlantic Africa,* 355–71.

Gaioso, Raimundo José de Sousa. *Compêndio histórico-político dos princípios da lavoura do Maranhão.* Paris: Rougeron, 1818.

Gamble, David P., and P.E.H. Hair, eds. *The Discovery of River Gambra (1623) by Richard Jobson.* London: Hakluyt Society, 1999.

Gaspar, David Barry. "From 'The Sense of Their Slavery': Slave Women and Resistance in Antigua, 1632–1763." In Gaspar and Hine, *More Than Chattel,* 218–38.

Gaspar, David Barry, and Darlene Clark Hine, eds. *More Than Chattel: Black Women and Slavery in the Americas.* Bloomington: Indiana University Press, 1996.

Gentry, Alwyn. "Diversity and Floristic Composition of Lowland Tropical Forest in Africa and South America." In Peter Goldblatt, ed., *Biological Relationships between Africa and South America,* 507–47. New Haven: Yale University Press, 1993.

Gepts, P. "A Comparison between Crop Domestication, Classical Plant Breeding, and Genetic Engineering." *Crop Science* 42 (2002): 1780–90.

Glick, Thomas F. *Islamic and Christian Spain in the Early Middle Ages.* Princeton: Princeton University Press, 1979.

Godinho, Vitorino Magalhães. *Os descobrimentos e a economia mundial.* 2 vols. Lisbon: Editora Arcádia, 1965.

Gray, Lewis C. *History of Agriculture in the Southern United States to 1860.* 2 vols. Gloucester, MA: Peter Smith, 1958.

Gremillion, Kristen. "Adoption of Old World Crops and Processes of Cultural Change in the Historic Southeast." *Southeastern Archaeology* 12, no. 1 (1993): 15–20.

Grimé, William Ed. *Ethno-botany of the Black Americans.* Algonac, MI: Reference Publications, 1979.

Grubben, G.J.M., and O.A. Denton, eds. *Plant Resources of Tropical Africa.* Vol. 2, *Vegetables.* Leiden: Backhuys, 2004.

Guimarães, Carlos Magno. "Esclavage, quilombos et archéologie." *Les Dossiers d'Archéologie,* no. 169 (1992): 67.

Hair, P.E.H., Adam Jones, and Robin Law. *Barbot on Guinea: The Writings of Jean Barbot on West Africa, 1678–1712.* 2 vols. London: Hakluyt Society, 1992.

Hall, Frederick, William F. Harrison, and Dorothy Winters Welker. *Dialogues of the Great Things of Brazil.* Albuquerque: University of New Mexico Press, 1987.

Hall, Robert L. "Savoring Africa in the New World." In H.J. Viola and C. Margolis, eds., *Seeds of Change,* 160–71. Washington, D.C.: Smithsonian Institution Press, 1991.

Handler, Jerome S. "The Amerindian Slave Population of Barbados in the Seventeenth and Early Eighteenth Centuries." *Caribbean Studies* 8, no. 4 (1969): 38–64.

———. "Plantation Slave Settlements in Barbados, 1650s to 1834." In Alvin O. Thompson, ed., *In the Shadow of the Plantation*, 123–61. Kingston, Jamaica: Ian Randle, 2002.

Handler, Jerome S., and Michael L. Tuite. "The Atlantic Slave Trade and Slave Life in the Americas: A Visual Record." Database project of Virginia Foundation for the Humanities and Digital Media Lab at the University of Virginia Library. http://hitchcock.itc.virginia.edu/Slavery/index.php.

Hanotte, Olivier, D. G. Bradley, J. W. Ochieng, Y. Verjee, E. W. Hill, and J. E. Rege. "African Pastoralism: Genetic Imprints of Origins and Migrations." *Science* 296, no. 5566 (2002): 336–39.

Hanson, A. A. *Grass Varieties in the United States.* Agriculture Handbook No. 170. Washington, D.C.: USDA, 1972.

Harlan, Jack R. "Agricultural Origins: Centers and Noncenters." *Science,* New Series 174, no. 4008 (1971): 468–74.

———. *Crops and Man.* Madison, WI: Crop Society of America, 1975.

———. "The Tropical African Cereals." In David R. Harris and Gordon C. Hillman, eds., *Foraging and Farming: The Evolution of Plant Exploitation,* 337–43. London: Unwin Hyman.

Harlan, Jack R., Jan M. J. de Wet, and Ann B. L. Stemler. *Origins of African Plant Domestication.* The Hague: Mouton, 1976.

Harms, Robert. *The Diligent: A Voyage through the Worlds of the Slave Trade.* New Haven: Yale University Press, 2002.

Harris, David R. *Plants, Animals, and Man in the Outer Leeward Islands, West Indies: An Ecological Study of Antigua, Barbuda, and Anguilla.* University of California Publications in Geography 18. Berkeley: University of California Press, 1965.

Harris, Jessica B. *The Africa Cookbook.* New York: Simon and Schuster, 1998.

———. *Iron Pots and Wooden Spoons.* New York: Atheneum Press, 1989.

Hartsinck, J. J. *Beschrijving van Guiana: Part II.* Amsterdam: S. Emmering, 1974 [1770].

Hauser, Mark W. "Between Urban and Rural: Organization and Distribution of Local Pottery in Eighteenth-Century Jamaica." In Ogundiran and Falola, *Archaeology of Atlantic Africa,* 292–310.

Hawthorne, Walter. *Planting Rice and Harvesting Slaves.* Portsmouth, NH: Heinemann, 2003.

Heath, Barbara J., and Amber Bennett, "'The little Spots allow'd them': The Archaeological Study of African-American Yards." *Historical Archaeology* 34, no. 2 (2000): 38–55.

Hemming, John. *Red Gold: The Conquest of the Brazilian Indians, 1500–1760.* Cambridge, MA: Harvard University Press, 1978.

Henderson, James. *A History of the Brazil: Comprising Its Geography, Commerce, Colonization, Aboriginal Inhabitants.* London: Longhurst, 1821.

Herbert, Eugenia. "Smallpox Inoculation in Africa." *Journal of African History* 16 (1975): 539–59.

Herlin, Susan J. "Brazil and the Commercialization of Kongo, 1840–1870." In Curto and Lovejoy, *Enslaving Connections,* 261–83.

Herskovits, Melville J., and Frances S. Herskovits. *Rebel Destiny: Among the Bush Negroes of Dutch Guiana.* New York: McGraw-Hill, 1934.

Hess, Karen. *The Carolina Rice Kitchen.* Columbia: University of South Carolina Press, 1992.

———. "Mr. Jefferson's Table: The Culinary Legacy of Monticello." Unpublished ms.

Heywood, Linda M., and John K. Thornton. *Central Africans, Atlantic Creoles, and the Foundation of the Americas, 1585–1660.* New York: Cambridge University Press, 2007.

Highfield, Arnold R. *J. L. Carstens' St. Thomas in Early Danish Times.* St. Croix: Virgin Islands Humanities Council, 1997.

Highfield, Arnold R., and Vladimir Barac. *C. G. A. Oldendorp's History of the Mission of the Evangelical Brethren on the Caribbean Islands of St. Thomas, St. Croix, and St. John.* Ann Arbor, MI: Karoma Publishers, 1987.

Higman, Barry. *Slave Populations of the British Caribbean, 1807–1834.* Kingston, Jamaica: University of the West Indies, 1995.

Hoehne, F. C. *Botánica e agricultura no Brasil no século XVI.* São Paulo: Companhia Editora Nacional, 1937.

Holl, Augustin F. C. *Saharan Rock Art: Archaeology of Tassilian Pastoralist Iconography.* New York: Altamira Press, 2004.

Howard, Patricia L. "Women and the Plant World: An Exploration." In Patricia L. Howard, ed., *Women and Plants: Gender Relations in Biodiversity Management and Conservation,* 1–48. London: Zed, 2003.

Howe, George. "Last Slave Ship." *Scribner's Magazine* 8, no. 1 (July 1890): 113–29.

Hughes, Griffith. *The Natural History of Barbados.* New York: Arno Press, 1972.

Humboldt, Alexander von. *Personal Narrative of Travels to the Equinoctial Regions of America, during the Years 1799–1804.* 3 vols. London: Henry G. Bohn, 1853.

Hurault, J. *La vie matérielle des noirs réfugiés Boni et des indiens Wayana du Haut-Moroni (Guyane Française): Agriculture économie et habitat.* Paris: ORSTOM, 1965.

Irvine, F. R. "The Edible Cultivated and Semi-cultivated Leaves of West Africa." *Qualitas Plantarum et Materiae Vegetabiles,* no. 2 (1956): 35–42.

———. "Supplementary and Emergency Food Plants of West Africa." *Economic Botany* 6 (1952): 23–40.

Jackson, S. Douglas, ed. *The Principal Navigations, Voyages, Traffiques and Discoveries of the English Nation by Richard Hakluyt.* 8 vols. London: J. M. Dent, 1907.

James, Louis. "Eating the Dead." In Hena Maes-Jelinek, Gordon Collier, Geoffrey V. Davis, eds. *A Talent(ed) Digger: Creations, Cameos, and Essays in Honour of Anna Rutherford,* 309–12. Atlanta: Rodopi, 1996.

Jeffreys, M. D. W. "How Ancient Is West African Maize?" *Africa* 33, no. 2 (1963): 115–31.

Jobson, Richard. *The Golden Trade*. Devonshire, England: Speight and Walpole, 1904 [1623].

Jones, Adam. *German Sources for West African History, 1599–1669*. Wiesbaden: Franz Steiner Verlag, 1983.

Jones, Adam, and P. E. H. Hair. "Sources on Early Sierra Leone: (11) Brun, 1624." *Africana Research Bulletin* 7, no. 3 (1977): 52–64.

Jones, W. O. "Manioc: An Example of Innovation in African Economies." *Economic Development and Cultural Change* 5, no. 2 (1957): 97–117.

Jordaan, Han. "The Curaçao Slave Market: from *Asiento* Trade to Free Trade, 1700–1730." In Postma and Enthoven, *Riches from Atlantic Commerce*, 219–57.

Jordan, Terry G. *Trails to Texas: Southern Roots of Western Cattle Ranching*. Lincoln: University of Nebraska Press, 1981.

Juhé-Beaulaton, Dominique. "La diffusion du maïs sur les Côtes de l'Or et des esclaves aux XVII et XVIII siècles." *Revue Française d'Historie d'Outre-mer* 77, no. 287 (1990): 177–98.

Karasch, Mary. "Slave Women on the Brazilian Frontier in the Nineteenth Century." In Gaspar and Hine, *More Than Chattel*, 79–96.

Kea, Ray A. "Expansions and Contractions: World-Historical Change and the Western Sudan World-System (1200/1000 B.C.–1200/1250 A.D.)." *Journal of World-Systems Research* 10, no. 3 (2004): 723–816.

———. *Settlements, Trade, and Polities in the Seventeenth-Century Gold Coast*. Baltimore: Johns Hopkins University Press, 1982.

Keys, David. "Kingdom of the Sands." *Archaeology* (March/April 2005): 26–29.

Kimber, Clarissa Thérèse. *Martinique Revisited: The Changing Plant Geographies of a West Indian Island*. College Station: Texas A&M University Press, 1988.

Kiple, Kenneth, and Kriemheld Ornelas, eds. *The Cambridge World History of Food*. 2 vols. Cambridge: Cambridge University Press, 2000.

Kleiman, Kairn. *The Pygmies Were Our Compass: Bantu and Batwa in the History of West Central Africa, Early Times to c. 1900 C.E.* Portsmouth, NH: Heinemann, 2003.

Klein, Herbert S. *The Atlantic Slave Trade*. Cambridge: Cambridge University Press, 1999.

———. "The Atlantic Slave Trade to 1650." In Schwartz, *Tropical Babylons*, 201–36.

Klein, Martin A. "The Slave Trade and Decentralized Societies." *Journal of African History* 42 (2001): 49–65.

———. "Women and Slavery in the Western Sudan." In Robertson and Klein, *Women and Slavery in Africa*, 67–88.

Klooster, Wim. *Illicit Riches: Dutch Trade in the Caribbean, 1648–1795*. Leiden: KITLV Press, 1998.

Knight, Frederick. "In an Ocean of Blue: West African Indigo Workers in the Atlantic World to 1800." In Michael A. Gomez, ed., *Diasporic Africa: A Reader,* 28–44. New York: New York University, 2006.

Krzyzaniak, Lech. "New Light on Early Food-Production in the Central Sudan." *Journal of African History* 19, no. 2 (1978): 159–72.

Kuper, Rudolf, and Stefan Kroepelin. "Climate-Controlled Holocene Occupation in the Sahara: Motor of Africa's Evolution." *Science* 313, no. 5788 (2006): 803–7.

Kupperman, Karen Ordahl. *Providence Island, 1630–1641: The Other Puritan Colony.* Cambridge: Cambridge University Press, 1993.

———. *Roanoke: the Abandoned Colony.* Savage, MD: Rowman and Littlefield, 1984.

Labat, Jean-Baptiste. *Nouveau voyage aux isles de l'Amerique.* 6 vols. Paris, 1742.

Lacoste, Yves. *Ibn Khaldun: The Birth of History and the Past of the Third World.* London: Verso, 1984.

Lamming, George. *In the Castle of My Skin.* London: Longman, 1970.

Lane, Kris. "Africans and Natives in the Mines of Spanish America." In Matthew Restall, ed. *Beyond Black and Red: African-Native Relations in Colonial Latin America,* 159–84. Albuquerque: University of New Mexico Press, 2005.

Langdon, Robert. "The Banana as a Key to Early American and Polynesian History." *Journal of Pacific History* 28 (1993): 15–35.

Lefler, Hugh Talmage, ed. *A New Voyage to Carolina by John Lawson.* Chapel Hill: University of North Carolina Press, 1967 [1709].

Lejju, B. Julius, P. Robertsha, and D. Taylor. "Africa's Earliest Bananas?" *Journal of Archaeological Science* 33 (2006): 102–13.

Lemos Coelho, Francisco de. *Description of the Coast of Guinea, 1684.* Translated by P. E. H. Hair. Liverpool: University of Liverpool, 1985.

Levtzion, N., and J. F. P. Hopkins. *Corpus of Early Arabic Sources for West African History.* Princeton, NJ: Marcus Wiener Publishers, 2000.

Lewicki, Tadeusz. *West African Food in the Middle Ages.* Cambridge: Cambridge University Press, 1974.

Lewis, Bernard. *Race and Color in Islam.* New York: Harper and Row, 1971.

Lhote, Henri. *The Search for the Tassili Frescoes.* New York: E. P. Dutton, 1959.

Lieberman, Philip, and Robert McCarthy. "Tracking the Evolution of Language and Speech: Comparing Vocal Tracts to Identify Speech Capabilities." *Expedition* 49, no. 2 (Summer 2007): 15–20.

Lier, R. A. J. van. *Frontier Society: A Social Analysis of the History of Surinam.* The Hague: Martinus Nijhoff, 1971.

Ligon, Richard. *A True and Exact History of the Island of Barbadoes.* London: Frank Cass & Co., 1970 [ca. 1647].

Littlefield, Daniel C. *Rice and Slaves.* Baton Rouge: Louisiana State University, 1981.

Livermore, Harold V. *Royal Commentaries of the Incas and General History of Peru.* 2 vols. Austin: University of Texas Press, 1966.

Long, Edward. *History of Jamaica,* 3 vols. London: Lowndes, 1774.

Lovejoy, Paul E. *Caravans of Kola: The Hausa Kola Trade, 1700–1900*. Zaria, Nigeria: Ahmadu Bello University Press: 1980.

———. *Transformations in Slavery*. Cambridge: Cambridge University Press, 2000.

Lydekker, Richard. *The Sheep and Its Cousins*. London: G. Allen, 1912.

MacDonald, Kevin C. "The Origins of African livestock: Indigenous or Imported?" In Blench and MacDonald, *Origins and Development of African Livestock*, 2–17.

MacNeish, Richard. *The Origins of Agriculture and Settled Life*. Norman: University of Oklahoma Press, 1992.

Mann, Charles C. *1491: New Revelations of the Americas before Columbus*. New York: Knopf, 2005.

Marcgrave, Jorge. *História natural do Brasil*. São Paulo: Imprensa Oficial do Estado, 1942.

Marcus, Jacob R., and Stanley F. Chyet, eds. *Historical Essay on the Colony of Surinam, 1788*. Cincinnati: American Jewish Archives, 1974.

Marees, Pieter de. *Description and Historical Account of the Gold Kingdom of Guinea (1602)*. Translated and edited by Albert van Dantzig and Adam Jones. Oxford: Oxford University Press, 1987.

Marin, D. H., T. B. Sutton, and K. R. Barker. "Dissemination of Bananas in Latin America and the Caribbean and Its Relationship to the Occurrence of *Radopholus similis*." *Plant Disease* 82, no. 9 (1998): 964–74.

Marks, Morton. "Exploring *El Monte*: Ethnobotany and the Afro-Cuban Science of the Concrete." http://ilarioba.tripod.com/scholars/mortonmarks.htm.

Marshall, R., and E. Hildebrand. "Cattle before Crops: The Beginnings of Food Production in Africa," *Journal of World Prehistory* 16, no. 2 (2002): 99–143.

Marshall, Woodville K. "Provision Ground and Plantation Labor in Four Windward Islands: Competition for Resources during Slavery." In Berlin and Morgan, *Cultivation and Culture*, 203–20.

Marsters, Kate F. *Travels in the Interior Districts of Africa by Mungo Park*. Durham, NC: Duke University Press, 2000.

Martin, Bernard, and Mark Spurrell. *The Journal of a Slave Trader (John Newton) 1750–1754*. London: Epworth Press, 1962.

Martin, Franklin W., Ruth M. Ruberté, and Laura S. Meitzner. *Edible Leaves of the Tropics*. North Fort Myers, FL: Echo Press, 1998.

Maxwell, Kenneth. *Conflicts and Conspiracies: Brazil and Portugal, 1750–1808*. New York: Routledge, 2004.

Mbida, C. M., W. Van Neer, H. Doutrelepont, and L. Vrydaghs. "Evidence for Banana Cultivation and Animal Husbandry During the First Millennium BC in the Forest of Southern Cameroon." *Journal of Archaeological Science* 27 (2000): 151–62.

McCann, James C. *Maize and Grace: Africa's Encounter with a New World Crop, 1500–2000*. Cambridge, Mass: Harvard University Press, 2005.

McClellan, James E., III. *Colonialism and Science: Saint Domingue in the Old Regime*. Baltimore: Johns Hopkins University Press, 1992.

McIntosh, Susan K. "Paleobotanical and Human Osteological Remains." In S. K. McIntosh, ed., *Excavations at Jenné Jeno, Hambarketolo, and Kaniana (Inland Niger Delta, Mali, the 1981 Season)*, 348–53. Berkeley: University of California Press, 1995.

McNeill, J. R. "Biological Exchange and Biological Invasion in World History." Paper presented at the 19th International Congress of the Historical Sciences, Oslo, August 6–13, 2000.

McWilliams, James E. *A Revolution in Eating: How the Quest for Food Shaped America*. New York: Columbia University Press, 2005.

Meghen, Ciaran, David MacHugh, B. Sauveroche, G. Kana, and Dan Bradley, "Characterization of the Kuri Cattle of Lake Chad Using Molecular Genetics Techniques." In Blench and MacDonald, *Origins and Development of African Livestock,* 259–68.

Mello e Souza, Laura de. "Violência e práticas culturais no cotidiano de uma expedição contra quilombolas: Minas Gerais, 1769." In João José Reis and Flavio dos Santos Gomes, eds., *Liberdade por um fio,* 193–212. São Paulo: Companhia das Letras, 1996.

Mercer, John. *Canary Islands: Fuerteventura*. Harrisburg, PA: Stackpole Books, 1973.

Miller, Joseph C. "Retention, Reinvention, and Remembering: Restoring Identities through Enslavement in Africa and under Slavery in Brazil." In Curto and Lovejoy, *Enslaving Connections,* 81–121.

———. *Way of Death: Merchant Capitalism in the Angolan Slave Trade, 1730–1830*. Madison: University of Wisconsin Press, 1999.

Mintz, Sidney W. "Slavery and the Rise of Peasantries," *Historical Reflections* 6 (Summer 1979): 213–42.

Mintz, Sidney W., and Douglas Hall, *The Origins of the Jamaican Internal Marketing System*. Yale University Publications in Anthropology 57. New Haven: Yale University Press, 1960.

Miracle, Marvin. *Maize in Tropical Africa*. Madison, Wisconsin: University of Wisconsin Press, 1966.

Mollien, Gaspard. *Travels in Africa*. London: Sir Richard Phillips and Co., 1820.

Mooney, Carolyn. "Anthropologist Sheds Light on Jungle Communities Founded by Fugitive Slaves." *Chronicle of Higher Education,* May 22, 1998, B2.

Moore, Francis. *Travels into the Inland Parts of Africa*. London: Edward Cave, 1738.

Morgan, Philip D. *Slave Counterpoint: Black Culture in the Eighteenth-Century Chesapeake and Lowcountry*. Chapel Hill: University of North Carolina Press, 1998.

———. "Work and Culture: The Task System and the World of Low Country Blacks, 1700 to 1880," *William and Mary Quarterly,* 3rd series, vol. 39, no. 4 (1982): 536–99.

Mors, Walter B., Carlos Toledo Rizzini, and Nuno Álvares Pereira. *Medicinal Plants of Brazil*. Algonac, MI: Reference Publications, 2000.

Mouser, Bruce L. *A Slaving Voyage to Africa and Jamaica: The Log of the Sandown, 1793–1794*. Bloomington: Indiana University Press, 2002.

———. "Who and Where Were the Baga? European Perceptions from 1793 to 1821." *History in Africa* 29 (2002): 337–64.

Mullin, Michael. *Africa in America*. Urbana: University of Illinois Press, 1994.

Muzzolini, Alfred. *L'Art rupestre préhistorique des massifs centraux sahariens*. Oxford: British Archaeological Reports, 1986.

Nash, Gary B. *Red, White and Black: The Peoples of Early North America*. Englewood Cliffs, NJ: Prentice-Hall, 1992.

National Research Council (NRC). *Lost Crops of Africa*. Vol. 1, *Grains*. Washington, D.C.: National Academy Press, 1996.

———. *Lost Crops of Africa*. Vol. 2, *Vegetables*. Washington, D.C.: National Academy Press, 2006.

Neumann, Katharina. "Early Plant Food Production in the West African Sahel: New Evidence." In Marijke Van der Veen, ed., *The Exploitation of Plant Resources in Ancient Africa*, 73–80. New York: Kluwer Academic, 1999.

Neumann, Katharina, A. Butler, and S. Kahlheber, eds. *Food, Fuel and Fields: Progress in African Archaeobotany*. Cologne: Heinrich-Barth-Institut, 2003.

Neville, H. O. "The Cattle Industry of Cuba." *Cuba Review and Bulletin* (1920): 13–44.

Newsom, Lee A., and Deborah M. Pearsall. "Trends in Caribbean Island Archaeobotany." In Paul E. Minnis, ed., *People and Plants in Ancient Eastern North America*, 347–412. Washington, D.C.: Smithsonian Books, 2003.

Newson, Linda A., and Susie Minchin. *From Capture to Sale: The Portuguese Slave Trade to Spanish South America in the Early Seventeenth Century*. Boston: Brill, 2007.

Nogueira, J. M. F., P. J. P. Fernandes, and A. M. D. Nascimento. "Composition of Volatiles of Banana Cultivars from Madeira Island." *Phytochemical Analysis* 14 (2003): 87–90.

Nørregård, Georg. *Danish Settlements in West Africa, 1658–1850*. Boston: Boston University Press, 1966.

Northrup, David. *Trade without Rulers: Pre-Colonial Economic Development in South-Eastern Nigeria*. Oxford: Oxford University Press, 1978.

Ogundiran, Akinwumi, and Toyin Falola, eds. *Archaeology of Atlantic Africa and the African Diaspora*. Bloomington: Indiana University Press, 2007.

Oldmixon, John. *The British Empire in America Containing the History of the Discovery, Settlement, Progress and State of the British Colonies of the Continent and Islands of America*. 2 vols. New York, Augustus M. Kelley, 1969 [1741].

Otto, J. S., and Nain E. Anderson. "Cattle Ranching in the Venezuelan Llanos and the Florida Flatwoods: A Problem in Comparative History." *Comparative Studies in Society and History* 28, no. 4 (1986): 672–83.

———. "The Origins of Southern Cattle-Grazing: A Problem in West Indian History." *Journal of Caribbean History* 21, no. 2 (1988): 138–53.

Oudschans Dentz, F. "De geschiedenis van de rijstbouw in Suriname." *Landbouwkundig Tijdschrift*, no. 691 (1944): 491–92.

Pace, Antonio, ed. *Luigi Castiglioni's "Viaggio: Travels in the United States of North America, 1785–1787."* Syracuse: Syracuse University Press, 1983.

Pacheco Pereira, Duarte. *Esmeraldo de situ orbis: Côte occidentale d'Afrique de Sud Maro-*

cain au Gabon. Translated by Raymond Mauny. Mémorias, no. 19. Bissau, Guinea-Bissau: Centro de Estudos da Guiné Portuguêsa, 1956.

Pares, Richard. *Merchants and Planters.* Cambridge: Cambridge University Press, 1960.

Parry, John H. "Plantation and Provision Ground." *Revista de historia de America* 39 (1955): 1–20.

Parsons, James J. "The 'Africanization' of the New World Tropical Grasslands," *Tübinger Geographische Studien* 34 (1970): 141–53.

———. "Spread of African Pasture Grasses to the American Tropics," *Journal of Range Management* 25 (1972): 12–17.

Patterson, Orlando. "Slavery and Slave Revolts: A Sociohistorical Analysis of the First Maroon War, 1665 to 1740." In R. Price, *Maroon Societies,* 243–92.

Pelling, Ruth. "Garamantian Agriculture and Its Significance in a Wider North African Context: The Evidence of the Plant Remains from the Fazzan Project." *Journal of North African Studies* 10, no. 3 (2005): 397–411.

Pérez, J., D. Albert, S. Rosete, L. Sotolongo, M. Fernandez, P. Delprete, and L. Raz. "Consideraciones etnobotánicas sobre el género Dioscorea *(Dioscoreaceae)* en Cuba." *Ecosistemas,* no. 2 (2005): 1–8.

Petersham, Maud, and Miska. *The Story Book of Rice.* Philadelphia: John C. Winston Co., 1936.

Pinckard, George. *Notes on the West Indies.* London: Longman, Hurst, Rees, and Orme, 1806.

Piso, Guilherme [Willem]. *História natural e médica da India ocidental.* Rio de Janeiro: Instituto Nacional do Livro, 1957 [1645].

Pitman, Frank W. "Slavery on the British West India Plantations." *Journal of Negro History* 11 (1926): 606–8.

Pollitzer, William S. *The Gullah People and Their African Heritage.* Athens: University of Georgia Press, 1999.

Popovic, Alexander. *The Revolt of African Slaves in Iraq in the 3rd/9th Century.* Princeton, NJ: Markus Wiener, 1999.

Porter, Charlotte. "Science at the Time of Columbus." In James R. McGovern, ed., *The World of Columbus,* 59–77. Macon, GA: Mercer University Press, 1992.

Porter, George Richardson. *The Nature and Properties of the Sugar Cane: With Practical Directions for the Improvement of Its Culture and the Manufacture of Its Products.* London: Smith, Elder, and Co., 1830.

Portères, Roland. "African Cereals: Eleusine, Fonio, Black Fonio, Teff, Brachiaria, Paspalum, Pennisetum, and African Rice." In Jack R. Harlan, Jan M. J. de Wet, and Ann B. L. Stemler, eds., *Origins of African Plant Domestication,* 409–52. The Hague: Mouton, 1976.

———. "Présence ancienne d'une varieté cultivée d'*Oryza glaberrima* en Guyane française." *Journal d'agriculture tropicale et de botanique appliquée* 11, no. 12 (1955): 680.

Postma, Johannes. *The Dutch in the Atlantic Slave Trade, 1600–1815*. Cambridge: Cambridge University Press, 1990.

———. "West-African Exports and the Dutch West India Company, 1675–1731." *Economisch- en Sociaal-Historisch Jaarboek* 36 (1973): 53–74.

Postma, Johannes, and Victor Enthoven, eds. *Riches from Atlantic Commerce*. Leiden: Brill, 2003.

Price, Richard. *First-Time: The Historical Vision of an Afro-American People*. Baltimore: Johns Hopkins University Press, 1983.

———, ed. *Maroon Societies: Rebel Slave Communities in the Americas*. Baltimore: Johns Hopkins University Press, 1979.

———. "Subsistence on the Plantation Periphery: Crops, Cooking, and Labour among Eighteenth-Century Suriname Maroons." *Slavery and Abolition*, 12, no. 1 (1991): 107–27.

Price, Richard, and Sally Price, eds. *Stedman's Surinam: Life in an Eighteenth-Century Slave Society*. Baltimore: Johns Hopkins University Press, 1992.

Price, Sally. *Co-wives and Calabashes*. Ann Arbor, MI: University of Michigan, 1993.

Pulsipher, Lydia M. "The Landscapes and Ideational Roles of Caribbean Slave Gardens." In N. Miller and K. L. Gleason, eds., *The Archaeology of Garden and Field*, 202–22. Philadelphia: University of Pennsylvania Press, 1994.

Rashford, John. "Arawak, Spanish and African Contributions to Jamaica's Settlement Vegetation." *Jamaica Journal* 24, no. 3 (1993): 17–23.

Reader, John. *Africa: A Biography of the Continent*. New York: Knopf, 1998.

Rebora, Giovanni, and Albert Sonnenfeld. *The Culture of the Fork*. New York: Columbia University Press, 2001.

Rediker, Marcus. *The Slave Ship: A Human History*. New York: Viking, 2007.

Reitz, Elizabeth J., and C. Margaret Scarry. *Reconstructing Historic Subsistence with an Example from Sixteenth-Century Spanish Florida*, Special Publication Series, no. 3. Glassboro, NJ: Society for Historical Archaeology, 1985.

Renkema, W. E. *Het Curaçaose plantagebedrijf in de negentiende eeuw*. Zutphen, The Netherlands: Walburg, 1981.

Reynolds, P. K. *The Banana: Its History, Cultivation and Place among Staple Foods*. New York: Houghton Mifflin, 1927.

———. "Earliest Evidence of Banana Culture." *Journal of the American Oriental Society Supplement*, no. 12 (1951): 1–28.

Ribeiro, Orlando. *Aspectos e problemas da expansão portuguésa*. Lisbon: Estudos de Ciencias Políticas e Sociais, Junta de Investigações do Ultramar, 1962.

Ribeiro, Ricardo Ferreira. *Florestas anãs do sertão: O cerrado na história de Minas Gerais*. Belo Horizonte, Brazil: Autêntica Editora, 2005.

Riley, Thomas J., Richard Edging, and Jack Rosen. "Cultigens in Prehistoric Eastern North America: Changing Paradigms." *Current Anthropology* 31, no. 5 (1990): 525–41.

Robertson, Claire, and Martin A. Klein. Introduction to Robertson and Klein, *Women and Slavery in Africa*.

————, eds. *Women and Slavery in Africa.* Madison: University of Wisconsin Press, 1983.

Rodney, Walter. "African Slavery and other Forms of Social Oppression on the Upper Guinea Coast in the Context of the Atlantic Slave Trade." In J. E. Inikori, ed., *Forced Migration,* 6–70. London: African Publishing Company, 1982.

————. *A History of the Upper Guinea Coast, 1545 to 1800.* New York: Monthly Review Press, 1970.

Roig, Juan Tomás. *Plantas medicinales, aromáticas or venenosas de Cuba.* 2 vols. Havana: Ed. Científico-Técninca, 1991, 1992 [1945].

Rossel, Stine, Fiona Marshall, Joris Peters, Tom Pilgram, Matthew D. Adams, and David O'Connor. "Domestication of the Donkey: Timing, Processes, and Indications." *Proceedings of the National Academy of Sciences (PNAS)* 105, no. 10 (2008): 3715–20.

Rouse, John E. *The Criollo: Spanish Cattle in the Americas.* Norman: University of Oklahoma Press, 1977.

Rowe, Phillip, and Franklin E. Rosales. "Bananas and Plantains." In Jules Janick and James N. Moore, eds., *Fruit Breeding,* vol. 1, *Tree and Tropical Fruits,* 167–211. New York: John Wiley, 1996.

Rugendas, Johann Moritz. *Viagem pitoresca através do Brasil.* São Paulo: Livraria Martins Editôra, 1954 [1821–25].

Rumeu de Armas, Antonio. *España en el Africa Atlántica.* 2 vols. Madrid: Instituto de Estudios Africanos, Consejo Superior de Investigaciónes Científicas, 1956–57.

Rutten, A. M. G. *Dutch Transatlantic Medicine Trade in the Eighteenth Century under the Cover of the West India Company.* Rotterdam: Erasmus, 2000.

Salaman, Redcliffe. *The History and Social Influence of the Potato.* Cambridge: Cambridge University Press, 1970.

Salley, A. S. "Introduction of Rice into South Carolina." *Bulletin of the Historical Commission of South Carolina* (Columbia), no. 6 (1919).

Sauer, Carl O. *The Spanish Main.* Berkeley: University of California Press, 1966.

Sauer, Jonathan D. *Historical Geography of Crop Plants.* Boca Raton, FL: CRC Press, 1993.

Saunders, A. C. de C. M. *A Social History of Black Slaves and Freedmen in Portugal, 1441–1555.* Cambridge: Cambridge University Press, 1982.

Schneider, Elizabeth. *Vegetables from Amaranth to Zucchini.* New York: William Morrow, 2001.

Schneider, John T. *Dictionary of African Borrowings in Brazilian Portuguese.* Hamburg: Helmut Buske Verlag, 1991.

Schoenbrun, David L. "Cattle Herds and Banana Gardens: The Historical Geography of the Western Great Lakes Region, ca AD 800–1500." *African Archaeological Review* 11 (1993): 39–72.

Schwartz, Stuart B. "The Mocambo: Slave Resistance in Colonial Bahia." In R. Price, *Maroon Societies,* 202–26.

———. "Resistance and Accommodation in Eighteenth Century Brazil: The Slaves' View of Slavery." *Hispanic American Historical Review* 57, no. 1 (1977): 69–81.

———, ed. *Tropical Babylons: Sugar and the Making of the Atlantic World, 1450–1680.* Chapel Hill: University of North Carolina Press, 2004.

Searing, James F. *West African Slavery and Atlantic Commerce: The Senegal River Valley, 1700–1860.* Cambridge: Cambridge University Press, 1993.

Sharp, William Frederick. *Slavery on the Spanish Frontier: The Colombian Chocó, 1680–1810.* Norman: University of Oklahoma Press, 1976.

Silveira, Simão Estácio da. *Relação sumária das cousas da Maranhão.* São Paulo: Editora Siciliano, 2001 [1624].

Simmonds, N. W. *Bananas.* London: Longmans, 1959.

Skelton, Raleigh Ashlin. *Magellan's Voyage: A Narrative Account of the First Navigation by Antonio Pigafetta.* New York: Dover, 1994.

Sloane, Hans. *A Voyage to the Islands Madera, Barbados, Nieves, San Christophers and Jamaica.* 2 vols. London: Printed by B. M. for the author, 1707–25.

Sluyter, Andrew. "The Role of Black Barbudans in the Establishment of Open-Range Cattle Herding in the Colonial Caribbean and South Carolina." *Journal of Historical Geography* 35 (2009): 330–49.

Smith, Andrew. *Peanuts: The Illustrious History of the Goober Pea.* Urbana: University of Illinois, 2002.

Smith, Ifeyironwa Francisca. *Food of West Africa: Their Origin and Use.* Ottawa: Kwik Kopy, 1998.

Soluri, John. *Banana Cultures.* Austin: University of Texas Press, 2005.

Sparks, Randy. "The Emergence of Afro-Atlantic Foodways." Paper presented at Cuisines of the Lowcountry and the Caribbean conference, Charleston, SC, March 20–23, 2003.

Spix, Johann Baptist von, and Carl Friedrich Philipp von Martius. *Travels in Brazil, in the Years 1817–1820.* 2 vols. London: Longman, Hurst, Rees, Orme, Brown, and Green, 1824.

Stahl, Anne Brower. "Early Food Production in West Africa: Rethinking the Role of the Kintampo Culture." *Current Anthropology* 27, no. 5 (1986): 532–36.

———. "Entangled Lives: The Archaeology of Daily Life in the Gold Coast Hinterlands, AD 1400–1900." In Ogundiran and Falola, *Archaeology of Atlantic Africa,* 49–76.

Stedman, John Gabriel. *Narrative, of a Five Years' Expedition, against the Revolted Negroes of Surinam, in Guiana.* 2 vols. London: J. Johnson, 1813.

Stephen, Leslie, and Sidney Lee, eds. *Dictionary of National Biography.* New York: Macmillan, 1893.

Stipriaan, A. van. *Surinaams Contrast.* Leiden: KITLV Uitgeverij, 1993.

Stover, R. H., and N. W. Simmonds. *Bananas.* London: Longmans, 1987.

Svalesen, Leif. *The Slave Ship Fredensborg.* Bloomington: Indiana University Press, 2000.

Sweet, James H. *Recreating Africa: Culture, Kinship, and Religion in the African-Portuguese World, 1441–1770.* Chapel Hill: University of North Carolina Press, 2003.

Syrett, Harold C., ed. *The Papers of Alexander Hamilton.* 27 vols. New York: Columbia University Press, 1963.

Taylor, John Martin. "Boiled Peanuts." *Gastronomica* 1, no. 4 (2001): 25–26.

Thomas, Hugh. *The Slave Trade: The Story of the Atlantic Slave Trade; 1440–1870.* New York: Simon and Schuster, 1999.

Thornton, John. *Africa and Africans in the Making of the Atlantic World, 1400–1680.* New York: Cambridge University Press, 1992.

Tomich, Dale. "Une Petite Guinée: Provision Ground and Plantation in Martinique, 1830–1848." In Berlin and Morgan, *Cultivation and Culture,* 221–42.

Toussaint-Samat, Maguelonne. *History of Food.* Malden, MA: Blackwell, 1999.

Trans-Atlantic Slave Trade Database Project. www.metascholar.org/TASTD-Voyages /index.html.

Tucker, Richard P. *Insatiable Appetite: The United States and the Ecological Degradation of the Tropical World.* Berkeley: University of California Press, 2000.

Turner, Lorenzo Dow. *Africanisms in the Gullah Dialect.* Columbia: University of South Carolina Press, 2002.

Unger, R. W. "Portuguese Shipbuilding and the Early Voyages to the Guinea Coast." In Felipe Fernández-Armesto, ed., *The European Opportunity,* vol. 2, *An Expanding World: The European Impact on World History, 1450–1800,* 43–63. Ashgate, England: Variorum, 1995.

Vaillant, A. "Milieu cultural et classification des variétés de riz des Guyanes français et hollandaise." *Revue Internationale de Botanique Appliquée et d'Agriculture Tropicale,* no. 33 (1948): 520–29.

Vansina, Jan. "Bananas in Cameroun, c. 500 BCE? Not Proven." *Azania* 38 (2004): 174–76.

———. *Paths in the Rainforest.* Madison: University of Wisconsin, 1990.

Vaughan, Duncan A., Bao-Rong Lu, and Norihiko Tomooka. "The Evolving Story of Rice Evolution." *Plant Science* 174 (2008): 394–408.

Vaughan, J. G., and C. A. Geissler. *The New Oxford Book of Food Plants.* Oxford: Oxford University Press, 1997.

Vieira, Alberto. "Sugar Islands: The Sugar Economy of Madeira and the Canaries, 1450–1650." In Schwartz, *Tropical Babylons,* 42–84, 57–58.

Viotti da Costa, Emilia. "The Portuguese-African Slave Trade: A Lesson in Colonialism." *Latin American Perspectives* 12 (1985): 41–61.

Visman, M. A. "Van slaaf tot plantagehouder: Een aspect van het 18e eeuws plantagewezen op Curaçao." *Nieuw West-Indische Gids* 55, nos. 1–2 (1981): 39–51.

Vlach, John Michael. *By the Work of Their Hands: Studies in Afro-American Folklife.* Charlottesville: University Press of Virginia, 1991.

Voeks, Robert A. *Sacred Leaves of Candomblé.* Austin: University of Texas Press, 1997.

Vogel, Arno, Marco Antonio da Silva Mello, and José Flávio Pessoa de Barros. *A galinha d'Angola: Iniciação e identidade na cultura Afro-Brasileira.* Rio de Janeiro: Universidade Federal Fluminense, 1993.

Vogt, John. *Portuguese Rule on the Gold Coast, 1469–1682.* Athens: University of Georgia Press, 1979.

Warner-Lewis, Maureen. *Guinea's Other Sons: The African Dynamic in Trinidad Culture.* Dover, MA: Majority Press, 1991.

Warren, George. *An Impartial Description of Surinam upon the Continent of Guiana in America with a History of Several Strange Beasts, Birds, Fishes, Serpents, Insects and Customs of That Colony, etc.* London: William Godbid for Nathaniel Brooke, 1667.

Watson, Andrew M. *Agricultural Innovation in the Early Islamic World.* Cambridge: Cambridge University Press, 1983.

Watson, William. "Some Account of an Oil, Transmitted by Mr. George Brownrigg, of North Carolina [October 31, 1769]." In *Philosophical Transactions of the Royal Society of London,* vol. 59, by the Royal Society (Great Britain), Charles Hutton, George Shaw, and Richard Pearson, 665–67. Published and printed by and for C. and R. Baldwin, 1809.

Watts, David. "Persistence and Change in the Vegetation of Oceanic Islands: An Example from Barbados, West Indies." *Canadian Geographer* 14, no. 1 (1970): 91–109.

———. *The West Indies: Patterns of Development, Culture and Environmental Change Since 1492.* Cambridge: Cambridge University Press, 1987.

Weber, Steven A. "Out of Africa: The Initial Impact of Millets in South Asia," *Current Anthropology* 39, no. 2 (1998): 267–274.

Wendorf, Fred, Angela Close, Romuald Schild, Krystyna Wasylikowa, Rupert Housley, Jack R. Harlan, and Halina Królik. "Saharan Exploitation of Plants 8,000 years BP." *Nature* 359, no. 6397 (1992): 721–24.

Wendorf, Fred, and Romuald Schild. "Nabta Playa and Its Role in Northeastern African Prehistory." *Journal of Anthropological Archaeology* 17 (1998): 97–123.

Wernicke, Edmundo. "Rutas y etapas de la introducción de los animales domésticos en las tierras americanas." *GAEA: Anales de la Sociedad Argentina de Estudios Geográficos* 6 (1938): 77–83.

West, Robert C. *The Pacific Lowlands of Colombia.* Baton Rouge: Louisiana State University Press, 1957.

Westergaard, Waldemar. *The Danish West Indies under Company Rule, 1675–1754.* New York: Macmillan, 1917.

Williamson, G., and W. J. A. Payne. *An Introduction to Animal Husbandry in the Tropics.* London: Longmans, 1960.

Williamson, Kay. "Food Plant Names in the Niger Delta." *International Journal of American Linguistics* 36, no. 2 (1970): 156–67.

Wilson, A. I., and D. J. Mattingly. "Irrigation Technologies: Foggaras, Wells and Field Systems." In D. J. Mattingly, C. M. Daniels, J. N. Dore, D. Edwards, and J. Hawthorn, eds., *The Archaeology of Fazzan,* vol. 1, *Synthesis,* 238–41. London: Society for Libyan Studies, 2003.

Wilson, Mary Tolford. "Peaceful Integration: The Owner's Adoption of His Slaves' Food." *Journal of Negro History* 49, no. 2 (1964): 116–27.

Wolf, Eric R. *Europe and the People without History.* Berkeley: University of California Press, 1982.

Wood, Peter H. *Black Majority.* New York: Knopf, 1974.

——. "The Calabash Estate: Gourds in African American Life and Thought." In *African Impact on the Material Culture of the Americas.* Winston-Salem, NC: Museum of Early Southern Decorative Arts, 1998.

——. "'It Was a Negro Taught Them': A New Look at African Labor in Early South Carolina," *Journal of Asian and African Studies* 9 (1974): 160–79.

Zaouali, Lilia. *Medieval Cuisine of the Islamic World.* Berkeley: University of California Press, 2007.

Zeder, Melinda A., Daniel G. Bradley, Eve Emshwiller, Bruce D. Smith, eds. *Documenting Domestication: New Genetic and Archaeological Paradigms.* Berkeley: University of California Press, 2006.

Zohary, Daniel, and Maria Hopf. *Domestication of Plants in the Old World.* Oxford: Oxford University Press, 2000.

INDEX

Page numbers in *italics* denote maps, illustrations, or photographs. The term *epi* in a reference to an endnote refers to information about the epigraph on the first page of the chapter (e.g. 187n*epi*).

highlands; savannas; savanna-to-forest
transitional zone
Ecuador, 103, 216n45
eddo, 220n37. *See also* taro
Edwards, Bryan, 167
eggplant (guinea squash), 191n42; Asian
  species, 191n42, 227n60; European
  accounts crediting slaves with diffusion of, 124, 125; and geographical
  references in plant names, 103; greens
  of, 178, 181n42; preparation of,
  224–25n31; savanna origin of, 22;
  slave ship subsistence and, 203n15;
  subsistence plots and, 135, 149, 227n60;
  uses of, 191n42
Elaeis guineensis. See oil palm (*dendê*)
elephant: forest, pre-Columbian exchanges
  and, 27, 193n1; ivory trade, 48
Eleusine coracana. See millet, finger
Elmina (Mina) fort: food preparation, 141,
  179–80, 223n17; markets and, 49, 50;
  population of, 49–50; soldier's rations,
  49. *See also* forts
El Salvador, 103
England: anti-slave trade enforcement by,
  202n8; dependence on African food
  surpluses, 49; guinea fowl and, 229n11;
  hide trade, 156; peace treaties with
  maroons, 206–7n3; and peanut oil,
  143, 224n29; privateering, 105, 107;
  Treaty of Breda and, 92, 210n35; water
  provision on slave ships, 69
Equus africanus, E. asinus. See donkey
Erythraean Sea, 33–34
Esteban, 89
ethnic specialist groups: flexible land-use
  practices encouraging, 25; knowledge
  of, and European settlement, 176;
  savannas and development of, 18
evolution of humans, in Africa, 8–9,
  188nn5–6
exchanges. *See* Columbian Exchange; pre-
  Columbian exchanges

farinha. *See* manioc
fava beans, 69, 203n14
Fernandes, Valentim, 41, 61, 63, 132, 156, 157

Ferreira, Alexandre Rodrigues, 128
findo. *See* fonio
flor de Jamaica. See hibiscuş
Florida, 89, 171, 218n4
fonio (funji, findo), 19, 22, 191n36; cuisine
  and, 180; Iberians encountering, *44*;
  Rome and, 194n13; slave ship subsistence and, 74; vernacular names for,
  19, 205n36
food. *See* African foodstaples; agriculture;
  Amerindian domesticates; shortages
  of food; subsistence; surpluses of food
food plots. *See* subsistence plots
food preparation techniques. *See* cuisines;
  milling; *specific foods*
foodstaples. *See* African foodstaples;
  Amerindian domesticates; beverages
forts, *52*; concentrations of, 49, *51*; food
  preparation, 141, 179–80, 223n17;
  livestock provisioning, 157; markets
  and, 49, 50; population of, 49–50;
  soldier's rations, 49; subsistence and,
  50, *52*, 54. *See also* military
France: laws regulating treatment of slaves,
  108–10, 126–27; subsistence plots,
  rights to, 133–34, 221n48; water provision on slave ships, 69
Fula people: cattle herds of, 156; and knowledge as herders, 172, 173–74; land-
  tenure practices of, 225n43; name of,
  233n66; as peasantry, 62–63, *62*; as
  slaves, 63–64
funerary offerings: ancient African traditions
  of, 94, 211n48; Creoles and yams and,
  211n47; maroon rice culture and, 93–94
funji. *See* fonio
Futa Jallon plateau: Bassari people and,
  187n*epi*; cattle domestication and, 12

Gambia River: and African rice, 22; and
  cattle herds, 156, 173; and greens,
  234n3; and kola nut, 70–71
Gamble, Samuel, 67
Garamantes people, 28–30, 165, 194n13
garden plots. *See* dooryard gardens; plantation provision grounds; subsistence
  plantations; subsistence plots

Kea, Ray, 49–50
Khaldun, Ibn, 31
Kikongo (Bantu language): Bambara
    groundnut and, 224n19; and slave
    escapes, 83, 88, 207n8
kitchen gardens. *See* dooryard gardens
kola nut, 72; commercial use of, 71, 204n26;
    origins of, 24; pre-Columbian ex-
    changese and, 31; religious use of, 71,
    238n36; and slave ship subsistence,
    70–72, 124; subsistence plots and, 135;
    vernacular names for, 71–72, 103, 141;
    water improvement with, 70–72, 125
Kongo, 47, 178, 181
Kupperman, Karen Ordahl, 104–5

Labat, Jean-Baptiste, 110, 157–58, 172
lablab (hyacinth bean, bonavist): area of
    domestication, 19; diffusion in Africa
    of, 19; edible portions of, 190–91n32;
    European accounts crediting slaves
    with diffusion of, 124; intercropped
    with sorghum, 147; plantation soci-
    eties and, 106; pre-Columbian
    exchanges, 19, 33; slave ship subsis-
    tence and, 69; subsistence plots and,
    135
Lablab purpureus. *See* lablab
Lagenaria siceraria. *See* gourd, bottleneck
Lamming, George, 237–38n35
land-use systems: grazing land held com-
    munally open, 225n43; pasture/
    croplands alternating, 25–26, 145,
    173–75
languages, shared, and slave escapes, 83, 88,
    207n8
Laurens, Henry, 130, 167, 169
Lavaux, Alexandre, 95–96, *95–96*
Lawson, John, 145, 149, 150–51, 152, 171
legumes: bean/pea linguistic distinction,
    192n51, 213n14; cuisine of rice and,
    182, 236n27; intercropping with, 24,
    25, 144, 147, 149; maroon communi-
    ties and, 88; pre-Columbian exchanges
    and, 27–28; provision plantations and,
    130; slave ship subsistence and, 67, 69,
    203n15; slave trade subsistence and,

52; soil benefits via, 192n51, 232n51;
    subsistence plots and, 128, 134, 145.
    See also cowpea; groundnuts; pigeon
    pea
Liberia, 48
Ligon, Richard, 106–7, 114–17, 161, 162,
    163–64, *164*, 230n29
Linnaeus, 90
Littleton, William, 68
livestock: African introductions to
    Americas, 105, 155–65, 229nn17,21,
    230nn31,34; alternating pastureland
    with crops, 25–26, 145, 173–75; Bar-
    buda and, 145, 160, 225n43; dooryard
    gardens and, 134; fodder, 116, 142,
    144, 149, 171, 225n35; goats, 13, 145;
    Iberian tradition, 172–73, 233n64;
    knowledge of African slaves, 172–76;
    maroon communities and, 84–85, 87;
    slave ship subsistence and, 15, 67, 116–
    17, 155–58, *158*; slave trade subsistence
    and, 52–53, *53*; tethering practices,
    173–75. *See also* cattle; domestication
    of animals and plants; donkey; guinea
    fowl; sheep
Louisiana, 182, 236n27
Lower Guinea Coast: maize introduction
    to, 55–59, *56*, 200nn25,28; okra and,
    140; yams and, 222n2
Lowndes, Thomas, 150
Luanda, 51, 157, 228n7, 234n3. *See also*
    Angola

Madeira Islands, sugar/slave trade and diffu-
    sion of plantains, 36–43, 195n29
maize: advantages of, 57–58; in Amerindian
    diet, 107; amounts produced, 200n25;
    cuisine and, 180, 181, 236n22; effect
    on African foodstaples, 55–56, *56*, 57–
    58; effect on agricultural production,
    55–56; maroon communities and, 88,
    97; milling of, 75; slave subsistence
    and, 126, 128; as slave trade enabler,
    55, 56–59; sorghum preferred over,
    144, 145; as symbol of slave status and
    sustenance, 55, 57
Mali, 22, 183, 183–84

tropics *(continued)*
America, 3, 102, 105, 107; yam developed as plant of, 24. *See also* African foodstaples; agriculture; grasses; rainforests, Old World
tubers. *See* root-crops
Tussac, François Richard de, 124

vegetable amaranth (African spinach/*bledo*/*callalou*): genera of Old and New World, use of, 179; maroon communities and, 99; pre-Columbian exchanges and, 193n2; religious use of, 238n36; vernacular names for, 235n16, 236n18
Venezuela, 169, 172
Vigna genus, 192n51
Vigna subterranea. See Bambara groundnut
Vigna unguiculata. See cowpea
Virginia, 126, 150, 170–71
Virgin Islands: cuisine and, 179; slave subsistence, 182; St. Croix, 117, 129, 130; St. Thomas, 108–9, 130

Warren, George, 109
Washington, Adelaide, *119*
watermelon: as African domesticate, 22; in candomblé religion, 91, 238n36; maroon communities and, 88, 94, 97; plantation societies and, 106
Windward Islands, subsistence plots and, 133

Wolof people, 63–64, 140, 236n27
women: as disporportionately retained, 61, 73, 201n35; dooryard gardens and, 134; gender ratios on slave ships, 71–73, 204n29; as greens experts, 177; as market vendors, 49, *50*, 181, *181*, 182–85, *183–84*, *Plate 8*; preparation of food by, slave trade and, 59, 60–61, 73–75, 201nn31,33, *Plate 3*; and rice among the maroons, 93–94. *See also* cuisines

yams: Amerindians and, 216n42; area of domestication, *17*, 24; in candomblé religion, 91; in cuisine, 180, 236n21; European accounts crediting slaves with diffusion of, 124; as funerary offering, 211n47; harvest festivals, 226n50; hurricane-prone areas and, 112, 118; maroon communities and, 88, 94, 97, 209n22; as medicinal, 24, 192n52; New World and Asian species of, 192n52, 216n42, 222n1; plantation societies and, 111–12; Portuguese encountering, *44*, 47; provision grounds and, 113, 129, 130; savanna origins of, 16, 24; slave trade and, 52, 57; subsistence plots and, 112–13, 128, 135, 145; timeframe for domestication, 16; toponyms for, 139–40; vernacular names for, 103, 113, 139–40, 222n2

Text: Garamond
Display: Garamond
Compositor: Integrated Composition Systems
Indexer: Victoria Baker
Cartographer: Bill Nelson
Printer and binder: Thomson-Shore